Cover Crops for Sustainable Farming

Eric Justes
Editor

Cover Crops for Sustainable Farming

Springer

Editor
Eric Justes
UMR AGIR (AGroécologie, Innovations & teRritoires) 1248
University of Toulouse, INRA, INPT, Centre de Recherches INRA Occitanie-Toulouse
Castanet Tolosan, France

UMR SYSTEM Univ Montpellier, Cirad, Inra, Montpellier SupAgro, CIHEAM-IAMM
Montpellier cedex 2, France

ISBN 978-94-024-0985-7 ISBN 978-94-024-0986-4 (eBook)
DOI 10.1007/978-94-024-0986-4

Library of Congress Control Number: 2017941216

Éditions Quæ, R10, 78026 Versailles cedex, France www.quae.com

Printed on acid-free paper

This Springer imprint is published by Springer Nature
The registered company is Springer Science+Business Media B.V.
The registered company address is: Van Godewijckstraat 30, 3311 GX Dordrecht, The Netherlands

The expert group would like to dedicate this study to the memory of Nadine Cohen-Brisson (INRA Avignon) and Fruck Dorsainvil (Faculty of Agronomics, Haiti), who sadly left us far too soon, in 2011 and 2010, respectively.

Nadine was the initial designer of the STICS crop model. She harnessed the support of her colleagues at the INRA as well as other French bodies to construct, parametrize and "validate" a French crop model, with its own specific characteristics. The model obtained the recognition of the international agronomic modeller community. For nearly 20 years, Nadine worked consistently to promote the generic nature and robustness of the STICS model, in a spirit of openness and continuous improvement of its performance and its validity range. Without her perspicacity, remarkable intelligence and insatiable and highly infectious joie de vivre, the simulation component of this study could not have been carried out – at least not in its current form and not with such a level of confidence in terms of the quality of the simulated results.

You have left behind a rich scientific heritage that the modeller community has a duty to promote. We owe you our greatest respect, Nadine, as a scientist and as a woman. We hope that we can be worthy of the commitment you demonstrated to French and international agronomics.

Fruck was a Haitian agronomic engineer who first came to France to prepare a thesis on the subject of catch crops at the Reims Agronomics Unit, supervised by Eric Justes and Jean-Marc Meynard (thesis supervisor). Fruck demonstrated immense courage, exceptional adaptability and incredible perspicacity when carrying out his research in a field that was completely unfamiliar to him. He tasted the joys of sampling frozen cover crops in champagne and learning the modelling trade. His work led to the development of specific formalisms for cover crops, with the complicit and ever-invaluable support of Nadine. When he returned to Haiti, Fruck worked to aid the agricultural development of the country he loved so much and to which he had decided to dedicate his enthusiasm and intelligence. You did not deserve to be swallowed up by the ground in the terrible earthquake that hit Haiti in 2010; you still had so much to offer to the agronomics of your country.

Nadine and Fruck, we will never forget your joie de vivre and your perspicacity!

Foreword

The presence of excess nitrate in surface water and groundwater poses problems in terms of public health and damage to the environment. First and foremost, it is a public health issue, particularly for groundwater; the water distributed for drinking water purposes must meet the required standard of containing less than 50 mg of nitrate per litre. It is also an environmental protection issue for surface water, especially for certain watersheds located close to the sea, as is the case with Brittany, for example; the rapid transport of nitrate from agricultural zones towards the sea leads to eutrophication of coastal zones.

Most of these nitrates come from agrosystems, due to the excessive use of nitrogen fertilizers but also to the natural production of nitrate by the mineralization of the organic matter contained in soil. The nitrate present in the soil is liable to be carried deeper into the ground, beyond the zone investigated, by the roots of crops, following the drainage of rainwater through the soil. This phenomenon is known as nitrate leaching. Furthermore, in temperate climates, the period of most intense drainage (autumn–winter) also corresponds to the period when the ground may be without any crop cover (bare soil), after a crop has been harvested in the summer or autumn and before the next crop is sown in the autumn or spring. It is therefore important to minimize the amount of nitrate present in the soil at the start of the drainage period in order to reduce leaching. The introduction of a cover crop traps some of this nitrate, which is then released into the soil in the weeks following destruction of the cover crop and is subsequently partially available for the next main crop.

This general principle of nitrate trapping during fallow periods is well known, but the dynamics of the nitrogen cycle processes and their precise quantification on a local scale are much more difficult to define. This is because the effectiveness of

The members of the expert group and the study project team, Justes, E.; Beaudoin, N.; Bertuzzi, P.; Charles, R.; Constantin, J.; Dürr, C.; Hermon, C.; Joannon, A.; Le Bas, C.; Mary, B.; Mignolet, C.; Montfort, F.; Ruiz, L.; Sarthou, J.P.; Souchère, V.; Tournebize, J.; Savini, I.; and Réchauchère, O., are indicated at the end of the book.

nitrate-trapping "catch crops" depends on a number of highly variable factors. The amount of nitrate present in the soil during fallow periods depends not only on farmers' fertilizing practices but also on the soil characteristics and the weather conditions that year. It is these factors that determine the rate at which organic nitrogen in the soil is biotransformed into nitrate, firstly by mineralization and secondly by nitrification. Moreover, the cover crop planting and destruction conditions, which determine their effectiveness, vary considerably and depend on local conditions as well as on the cropping systems. These conditions may create problems from a technical point of view and with respect to the organization of labour. The type of cover crop used is also very important in terms of trapping efficacy, linked to soil and climatic conditions.

The technical references available to the public authorities to help public policy-making are based on technical reviews produced by the CORPEN (the French steering committee for environmentally friendly farming practices), COMIFER (the French committee for the development of rational fertilization) and ANDA (the French National Association for Agricultural Development) about 10 years ago, as well as a few more recent publications by technical institutes and research bodies. To prepare for the fifth action programme (beginning in 2013), the public authorities (the French Ministries for Agriculture and Ecology) have requested an updated body of information and agronomic data in order to be able to propose measures that are as effective as possible to control nitrate pollution in surface water and groundwater.

The main objective of this study is to review, as objectively and exhaustively as possible, the available knowledge on nitrogen management during fallow periods using cover crops, for arable cropping systems employed in a range of soil and climate conditions across France, as well as their impact on the next crop. In their commissioning letter, the French Ministries for Agriculture and Ecology tasked the INRA (*Institut scientifique de recherche agronomique publique* – French National Institute for Agricultural Research) with conducting "a review of the available knowledge on nitrogen management during fallow periods, specifying the knowledge and research, as well as the uncertainties and questions requiring further investigation, with a view to extending the use of autumn cover crops". Anticipating that it would be impossible to achieve this objective simply by examining the scientific literature available for a range of soil and climate conditions making up the diversity of contexts found in mainland France, the contracting bodies accepted the INRA's proposal to supplement this review of knowledge based on an analysis of the international and French technical literature with a simulation study. The aim of the simulation study is to establish benchmarks for all French situations representative of arable cropping systems, including those that have not been studied by research work conducted on the basis of field experiments and monitoring. It also examines methods of nitrogen management during fallow periods other than cover crops (volunteers and crop residue management).

Furthermore, cover crops do not play only a nitrate-trapping role. They have other effects, which may be either positive or negative on the scale of the agrosystem. These are examined here through a review of the scientific literature.

This document therefore constitutes a synopsis of the main conclusions emerging from the analysis of the literature and the simulation studies, aimed at decision-makers and stakeholders: agricultural practitioners (farmers and advisers) and water and environmental managers. This synopsis is designed to be a key providing access to the various chapters of the study report. It closely follows the plan of the full report and makes frequent reference to it in the form of footnotes. This document first sums up (Sect. 2) current knowledge of the main agronomic and ecological functions of cover crops, determined on the basis of an analysis of the scientific and technical literature (without citing the bibliographic references, which are all listed in the report). It then summarizes (Sect. 3) the main results of the simulation studies aimed at representing different cover crop management methods for a broad range of soil and climate conditions representative of mainland France. The interim results or analyses explaining the main results presented here, along with other additional results, are detailed in the study report.

Contents

1 Study Context and Methodology 1
Eric Justes, Julie Constantin, Caroline Dürr, Carole Hermon, Christine Le Bas, Catherine Mignolet, Isabelle Savini, and Olivier Réchauchère

1.1 Fallow Period: Definition, Management Practices and Regulatory Aspects 1
1.1.1 Definitions 1
1.1.2 Cropping Practices 3
1.1.3 Regulatory Aspects 6
1.1.4 Requests for Exemption of Planting of Cover Crops 8
1.2 Study Objectives and Approach 9
1.2.1 Study Objectives 9
1.2.2 Study Approach and Organization 10

2 Main Lessons Drawn from the Analysis of the Literature 13
Eric Justes, Nicolas Beaudoin, Patrick Bertuzzi, Raphaël Charles, Julie Constantin, Carolyne Dürr, Alexandre Joannon, Christine Le Bas, Bruno Mary, Françoise Montfort, Laurent Ruiz, Jean-Pierre Sarthou, Véronique Souchère, and Julien Tournebize

2.1 Cover Crop Management Techniques 13
2.1.1 Cover Crop Planting Techniques 15
2.1.2 Cover Crop Destruction Techniques 17
2.1.3 Constraints in Terms of Organization of Labour and Economic Constraints on a Field Scale 18
2.2 Nitrate and Water During Fallow Periods 19
2.2.1 Soil Nitrogen, the Origin of Nitrate and Leaching 19
2.2.2 Nitrate Trapping Using Catch Crops 22
2.2.3 Effects Following Destruction of the Cover Crop 25
2.2.4 Cover Crops Within the Cropping System 26
2.3 Effects of Cover Crops on Erosion, Physical Soil Properties and Carbon Balance 27

2.3.1 Effects of Cover Crops on Erosion and Physical Soil Properties ... 27
2.3.2 Effect of Cover Crops on Carbon and Nitrogen Sequestration and Greenhouse Gas Emissions ... 29
2.4 Biotic Effects of Cover Crops on Weeds, Microflora and Fauna ... 30
2.4.1 Weeds and Fallow Period Management ... 30
2.4.2 Microflora and Fallow Period Management ... 33
2.4.3 Fauna (Invertebrates, Vertebrates) and Fallow Period Management ... 35
2.4.4 The Difficulties Related to Integrating the Various Biotic Effects ... 37
2.5 Conclusion: A Rich But Incomplete Analysis of the Literature ... 38
References ... 39

3 Main Lessons Drawn from the Simulation Study ... 41
Eric Justes, Julie Constantin, Nicolas Beaudoin, Carolyne Dürr, Bruno Mary, and Laurent Ruiz
3.1 Objectives of the Simulation Study Using a Dynamic Crop Model ... 41
3.2 Methodology of the Three Parts of the Simulation Study ... 43
3.2.1 The Situations Simulated ... 43
3.2.2 The Crop Model Chosen: STICS (Version 6.9) ... 47
3.2.3 Evaluation of Days Potentially Available in the Autumn for Destruction of the Cover Crop ... 49
3.2.4 Catch Crop Sowing-Emergence Phase Simulated Using the SIMPLE Model ... 49
3.2.5 Analysis of the Simulations and Presentation of Results ... 51
3.3 Generic and Specific Lessons Drawn from the Simulation Study ... 53
3.4 Discussion of Study Results ... 70
3.4.1 What Is the Validity Range of the Typical Rotations Simulated? ... 70
3.4.2 Can Catch Crops Mitigate Nitrate Leaching Situations with Excess Mineral Nitrogen in the Soil at the Time of Harvest? ... 71
3.4.3 An Approach for Defining the "Optimal" Date of Catch Crop Emergence and Destruction by Multi-criteria Analysis of Simulation Results ... 73
3.4.4 Cover Crops: What Other Ecosystem Services Can They Provide? ... 76
3.4.5 How Can Catch Crops Be Adapted Locally to Allow Their Broadest Possible Integration into Current Cropping Systems? ... 79
References ... 80

4 Conclusions and Outlook 83
Eric Justes
4.1 The Contribution of the Study: A Complement to the Analysis of the Literature to Help Quantify the Effects of Catch Crops and Fallow Period Management Methods in French Soil and Climate Conditions 83
4.2 Remaining Crucial Research Questions to be Explored 85

Erratum E1

Bibliography 89
Cropping Practices and Regulatory Aspects 89
Cover Crop Management Techniques 90
Nitrate and Water During Fallow Periods 90
Effects of Cover Crops on Erosion, Physical Soil Properties and Carbon Balance 92
Biotic Effects of Cover Crops on Weeds, Microflora and Fauna 92
Methodology of the Simulation Study 94
Authors 95
Driving of the Project 95

The original version of this book was revised.
An erratum to this book can be found at https://doi.org/10.1007/978-94-024-0986-4_5

Chapter 1
Study Context and Methodology

Eric Justes, Julie Constantin, Caroline Dürr, Carole Hermon, Christine Le Bas, Catherine Mignolet, Isabelle Savini, and Olivier Réchauchère

1.1 Fallow Period: Definition, Management Practices and Regulatory Aspects

1.1.1 Definitions

In crop rotation, the fallow period is the period between the harvest of the main crop (sown for the purpose of harvesting) and the sowing of the next (Fig. 1.1). Depending on the harvesting and sowing dates of main crops, it can last from several days in the case of late harvest followed by sowing of a winter crop, to several months (up to 9 months) in the case of a spring crop. During this period, soil that is left "bare" (without plants), especially in the event of a long fallow period, can significantly

The original version of this chapter was revised.
An erratum to this chapter can be found at https://doi.org/10.1007/978-94-024-0986-4_5

E. Justes (✉)
UMR AGIR (AGroécologie, Innovations & teRritoires) 1248, University of Toulouse, INRA, INPT, Centre de Recherches INRA Occitanie-Toulouse, 24 Chemin de Borde Rouge, Auzeville, CS 52627, Castanet Tolosan 31326, France

UMR SYSTEM Univ Montpellier, Cirad, Inra, Montpellier SupAgro, CIHEAM-IAMM, 2, place Pierre Viala, Bat. 27 , Montpellier 34060 Montpellier cedex 2, France
e-mail: eric.justes@inra.fr; eric.justes@cirad.fr

J. Constantin
UMR AGIR (AGroécologie, Innovations & teRritoires) 1248, University of Toulouse, INRA, INPT, Centre de Recherches INRA Occitanie-Toulouse, 24 Chemin de Borde Rouge, Auzeville, 31326 Castanet Tolosan cedex, Toulouse CS 52627, France
e-mail: julie.constantin@inra.fr

C. Dürr
INRA, UMR 1345 IRHS, 42 rue George Morel, Beaucouzé 49071, France
e-mail: carolyne.durr@inra.fr

E. Justes (ed.), *Cover Crops for Sustainable Farming*,
DOI 10.1007/978-94-024-0986-4_1

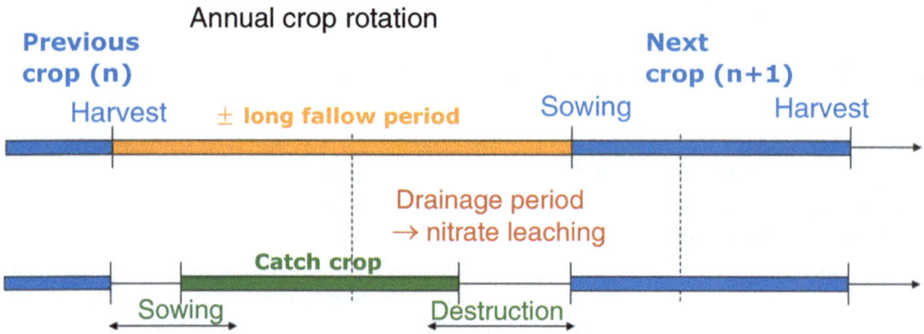

Fig. 1.1 Diagram showing the fallow period in an annual crop rotation (*top*) and the planting of a cover crop during this period (*bottom*)

increase the risk of leaching of nitrate into aquifers. The use of a cover crop[1] as a nitrate-trapping "catch crop" can reduce this phenomenon of nitrate transfer in variable proportions depending on the soil and climate conditions, as well as the cropping system. The reduction lies in the range of 20–90% compared to bare soil. This broad range of efficacy needs to be analysed on the basis of factors determining water drainage and nitrate leaching, as well as the other multiple functions of cover crops.

A cover crop is planted between two main crops, the aim being to provide a number of ecosystem services (agronomic and ecological) via agro-ecological functions. Primarily these may be: (i) to reduce leaching, (ii) to provide nitrogen to the next crop, (iii) to reduce erosion, (iv) to improve soil structure and soil hydric properties (v) to reduce parasite pressure on crops, (vi) to prevent weed growth, (vii) to increase the biodiversity of the farming landscape and environment (wildlife, bees, etc.). They are not destined to be harvested and are destroyed (or their growth is

[1] For vocabulary explanations, see the report, Chap. 1, Sect. 1.3.1.

C. Hermon
IEJUC (Institut des Etudes Juridiques de l'Urbanisme, de la Construction et de l'environnement (IEJUC), Université Toulouse Capitole, 2 rue Sainte Augustine, Toulouse 31500, France
e-mail: carole.hermon@ut-capitole.fr

C. Le Bas
INRA, US InfoSol, 2163 Avenue de la pomme de pin, Ardon, CS 40001, Orléans cedex 2 45075, France
e-mail: christine.le-bas@inra.fr

C. Mignolet
INRA, UR ASTER, 662 avenue Louis Buffet, Mirecourt 88500, France
e-mail: catherine.mignolet@inra.fr

I. Savini • O. Réchauchère
INRA, Unité Délégation aux Expertises, Prospectives et Etudes, 147 rue de l'Université, Paris 75338 PARIS Cedex 07, France
e-mail: isabelle.savini@inra.fr; olivier.rechauchere@inra.fr

stopped) before the next main crop is planted. Their biomass is returned to the soil to promote recycling of nutritional components for the next crop and improve the physical, chemical and biological fertility of the soil.

When a cover crop is planted with the primary or initial aim of reducing leaching in situations where the risk of such an occurrence is high, the term "catch crop" is used. When the primary objective is to supply nitrogen to the next crop – hence mainly in low input contexts – the term "green manure" is used. When the main aim is to cover the soil to limit water erosion and/or play a soil structuring role via the action of the roots, the term "cover crop" is used. It goes without saying that catch crops can fulfil several different agro-ecological functions simultaneously.

Cover crops – irrespective of their objectives – differ from secondary crops planted between two main crops in that the purpose of the latter is to produce a cash crop (grain or fodder) within a short period of time. The sole aim of cover crops, in contrast, is to provide agricultural or ecological services, with no direct benefit expected from the biomass produced considering this is not a cash crop even for bioenergy production.

1.1.2 Cropping Practices

Cover crops are usually sown after the previous main crop has been harvested, although they may also be undersown with the main crop. After harvest, these cover crops are generally sown between late July and early September and destroyed between November and February of the following year. Their growth period therefore ranges from 2 to 6 months, depending on crop rotations and regions. They may be destroyed naturally by frost, mechanically (chopping, ploughing, surface stubble ploughing) or chemically, by the application of a systemic foliar herbicide (glyphosate, for example), depending on the type of crop and the maturity of the plant cover, but also the regulations in force, which, in numerous cases, prohibit chemical destruction processes.

Current fallow period management practices in France were studied on the basis of information obtained from the *"Pratiques culturales en grandes cultures"* ("Arable cropping practices) survey conducted in 2006 by the French Ministry of Agriculture's Department of Statistics and Forward Studies. Since the results of the last survey conducted in 2011 were not yet available, this review of practices does not take into account the impact of the 2008 nitrates directive relative to autumn soil cover in "Nitrate vulnerable zones" (NVZs).

In the sample covered by the 2006 survey, 7.8 % of fields surveyed were planted with a catch crop and 20 % had previous crop volunteers. The majority of cover crops are Brassicaceae (66 %), with a large proportion of white mustard, which has "nematode control" properties, and 25 % are Poaceae; the use of mixed crops or other botanical families is not common (4–5 %). Soil cover practices during fallow periods are revealed to be highly diversified, in terms of extent (0–20 %), the cropping practice methods used to plant and destroy these cover crops (volunteers and catch crops), the region and previous/next main crop combinations. Regarding destruction, the survey indicates that mechanical chopping and/or burying are the main methods used in France.

The decision to use a cover crop depends on the crop that succeeds it (Fig. 1.2a): hence in 2006, for example, 48 % of sugar beet crops, 37 % of potato crops, 21 % of spring pea crops and 14 % of maize crops were preceded by a cover crop, representing a very marked increase from 2001, when the respective percentages were 21, 18, 4 and 5 % (Agreste Primeur, 2004). However, the planting of a cover crop still only happens in 4.5 % of fields before sunflowers (1 % in 2001) and in less than 4 % of fields before spring barley.

The great majority of cover crops follow straw cereal crops (13 % of wheat crops, 10 % of barley crops and 11 % of other cereal crops are followed by a cover crop) and, to a lesser extent, silage maize crops (6 % of maize crops) (Fig. 1.2b).

The practice of allowing previous crop volunteers to grow after harvest remained stable overall between 2001 (18 % of annual crops) and 2006 (20 %), although a few changes occurred. The previous crops concerned are primarily oilseed rape (volunteers are present in almost half of the fields in which oilseed rape crops have been grown), then straw cereals (29 % of fields with a previous durum wheat crop, 27 % of those with a previous common wheat or barley crop and 21 % of those with a previous other cereal) and protein crops (26 % of fields with a previous protein crop) (Fig. 1.2b).

An Initial Typology of Cropping Practices

Statistical analysis of the sample surveyed in 2006 led to the definition of eight main types of cover crop management. These eight management types (Fig. 1.3) correspond to cropping practice combinations that differ depending on the region and the main type of production system (arable, mixed arable/livestock farming or livestock farming).

Overall, the major results are as follows:

- Type 4, which accounts for 17 % of fields in the sample, is characteristic of cover crop management in an arable context in the Northern half of France: planting of mustard following wheat or barley, between 15 August and 15 September, destroyed mechanically between 1 November and 15 December, without organic manuring, before sowing a sugar beet, pea, grain maize or spring barley crop.
- Types 1–3, which are characteristic of both arable and mixed arable/livestock contexts, differ from type 4 as organic products are applied before the cover crop is sown (type 2) or at the time of sowing (types 1 and 3). They differ in terms of the cover crop sowing period (earlier for type 3), the type of crop sown or the kind of organic matter applied, and in terms of the cover crop destruction period (later and variable for type 1).
- For types 5–8, organic fertilizer is often applied,at the time of destruction or shortly after destruction of the cover crop. These inputs therefore do not concern the cover crop but, rather, the next crop. The proportion of fields concerned by this practice is different in each type.

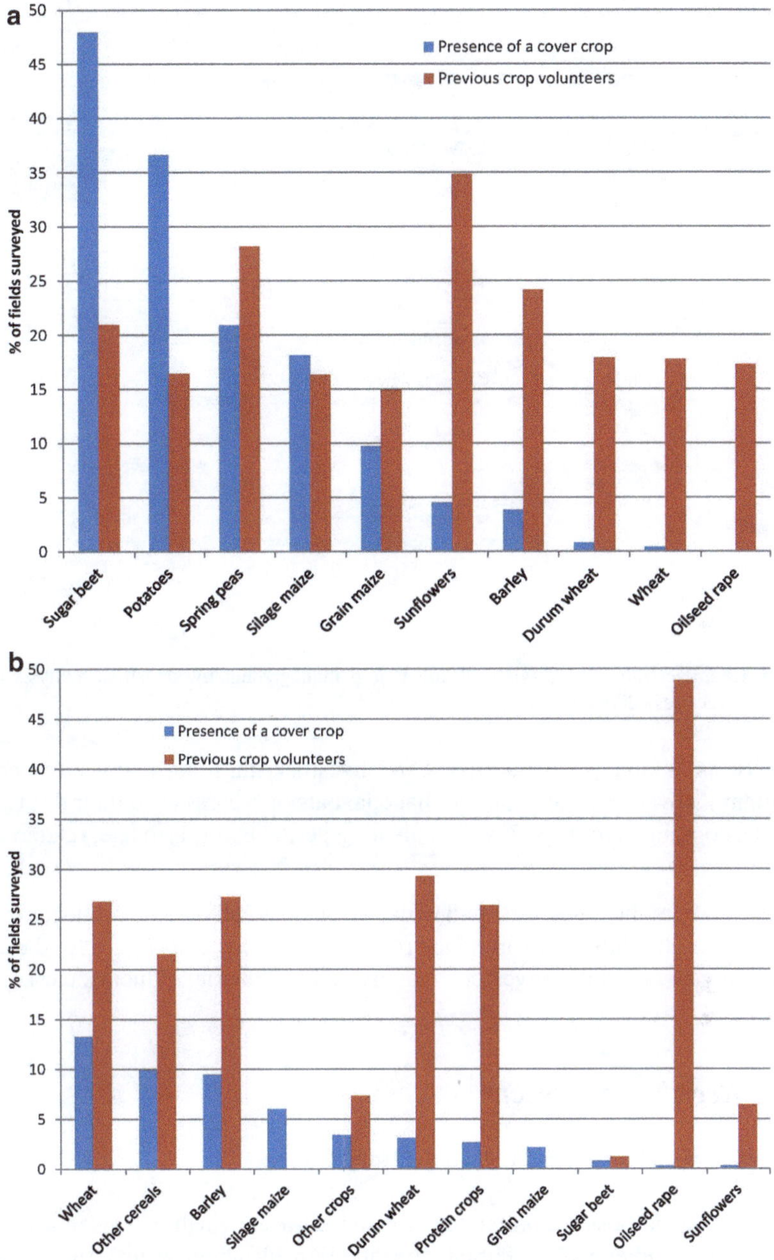

Fig. 1.2 Percentage of fields surveyed planted with a cover crop or with previous crop volunteers on the basis of (**a**) the next crop and (**b**) the previous crop ("Arable cropping practices", 2006 survey)

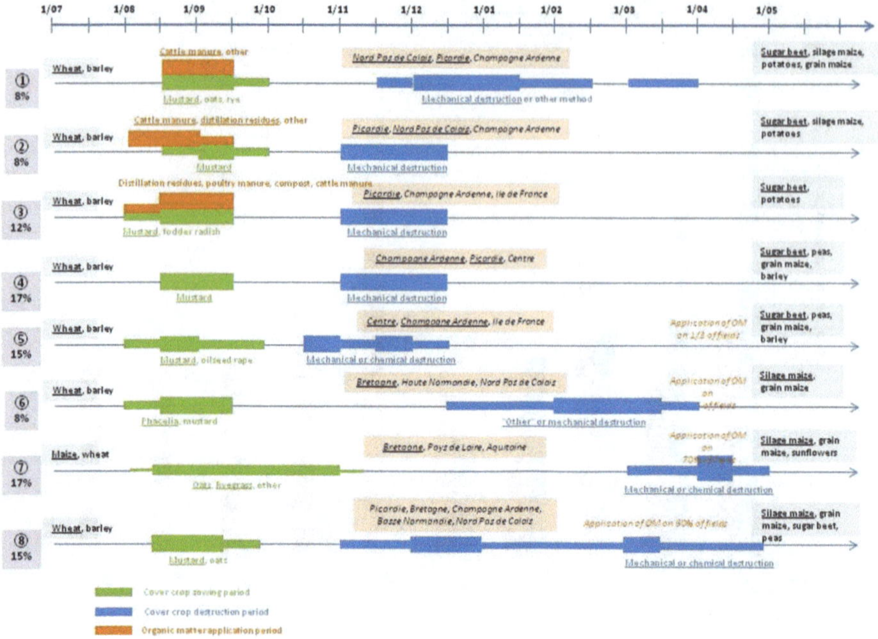

Fig. 1.3 Characterization of 8 types of cover crop management by statistical analysis of the "Cropping practices" 2006 survey

- For types 6 and 7, characteristic of livestock farming regions (especially Brittany), cover crops are mainly Phacelia, oats or ryegrass and their destruction is late, combined, for type 7, with a planting period that is also late, related to the previous maize crop.

Hence 28 % of the fields surveyed with a catch crop had organic products applied (types 1–3), highlighting the significant role of catch crops as a surface for the application of organic products (livestock manure, agro-industrial effluents, etc.).

1.1.3 Regulatory Aspects[2]

Main Features of the Legal and Regulatory Context

In 1975, for the purposes of protecting human health and via directives related to the quality of water intended for human consumption (drinking water), the European Community imposed maximum values for a variety of parameters. For example, the "nitrate" parameter is associated with a maximum limit of 50 mgNO_3/l, it being

[2] For a more detailed analysis of the legal and regulatory provisions related to soil cover requirements and their implementation, see the report, Chap. 2, Sect. 2.1.

accepted that it is the responsibility of each European Community Member State to decide on the *ad hoc* control policies to be implemented to obtain this result.

Furthermore, the "Nitrates" directive of 12 December 1991 stipulates a set of measures to be transposed in order to control leaching of nitrates from agricultural land. Member States must, firstly, identify "vulnerable zones" in their territories, on the basis of the criteria stipulated by the directive and, secondly, adopt and implement an action programme in these zones designed to reduce water pollution. The minimum content of these programmes is determined by the directive. This directive therefore sets an objective for the actions to be implemented to achieve the desired result.

More generally and more recently, in addition to the "nitrate" parameter itself, the Water Framework Directive (WFD) sets an objective for the good ecological and/or chemical status of water to be achieved by 2015, thereby clearly stipulating an obligation for results, except for exemptions strictly controlled by article 4.4 and 4.5 of the directive.

The Soil Cover Obligation in Action Programmes, As An Application of the Nitrates Directive

The legislative framework governing the first French Nitrates action programme (1996) did not incorporate as it should have done the provisions relative to ground cover contained in the "Code of good agricultural practices". This code recommends: "for annual cropping systems, improving the succession of crops in rotation so as to reduce the surface area of bare soil during periods presenting a risk of leaching, increasing the winter crop period compared to the spring crop period in the cropping plan, planting catch crops after crops that leave the soil bare and with a high mineral nitrogen content during long rainy periods".

The obligation to maintain winter soil cover was introduced for the first time in the 2nd Nitrates action programmes in 2001, in the context of "additional action zones", and it was then by a circular that the autumn soil cover of all "vulnerable zones" was recommended in 2008 (4th action programmes). These two dates are important stages in the construction of the policy package. However, the regulations all share one point in common: their application has been variable from one French *département* to the next, raising questions with respect to their consistency and legal legitimacy.

Regarding the 5th action programmes – currently in construction – the decree of 10 October 2011 includes fallow period cover crops in the national action plan. The order published on 7 May 2012 indicates that, in regional programmes, the mode of management of cover crops during fallow period will be specified in a later order scheduled for the end of 2012.

The Soil Cover Obligation Outside Action Programmes

In the context of the application of the French "Law on water and aquatic environments" (article L. 211-3 C. amended in 2006), the obligation to provide plant cover of the soil, either permanent or temporary, is one of the measures that may be enforced to ensure the protection of "areas draining into drinking water catchment zones of particular importance for current or future supplies".

In parallel, the second generation of *Schémas directeurs d'aménagement et de gestion des eaux* (SDAGEs – master plans for development and water management), adopted in late November/early December 2009 depending on the catchments, should ensure that the objective of the good ecological status of water set by the European water framework directive is met. Consequently, the SDAGEs may, independently of the Nitrates Directive transposition process, impose fallow period soil cover measures in order to achieve the objectives of the WFD.

Finally, the planning act of 3 August 2009 relating to the implementation of the French Environment Round Table, known as the Grenelle 1 law, targets, among all the other actions to be implemented by the State "the wider-spread use of cover crops in winter depending on local conditions" (recommended in the chapter dedicated to agriculture).

1.1.4 Requests for Exemption of Planting of Cover Crops

Following the publication of the circular extending autumn cover crops to all "vulnerable zones" recommended in 2008 (4th action programme), numerous applications for exemption have been submitted by farmer organisations and granted by the regional authorities. The arguments used in exemptions have been analysed and ranked by order of importance. The experts in charge of the simulations in the present study used this analysis to ensure that the scenarios tested adequately took into account situations in which farmers expect the implementation of catch crops to pose specific difficulties.

The first type of exemption concerns situations of soil with a high clay content. It relies on the argument that clay soils require early autumn tilling, something that is not easily compatible with the planting of a cover crop. However, it is worth noting that the texture of soils described as "clay" soils in these exemptions is highly heterogeneous with values varying from a clay content of 22–40 % depending on the *département*. If we refer to the *ad hoc* description of soil textures (GEPPA texture triangle widely used in France), a soil is classified as a clay soil when its clay content (fraction < 2 μm) is greater than 37 %, and as a clay-loam soil when it is between 27.5 and 37 %.

Other exemptions have been justified by: (i) low mineral nitrogen residual contents in the soil at harvest, (ii) late harvesting of the main crop (with exemption dates ranging from 1 September to 31 October depending on the *département*), not permitting planting of a cover crop, (iii) substitution of catch crops with cereal vol-

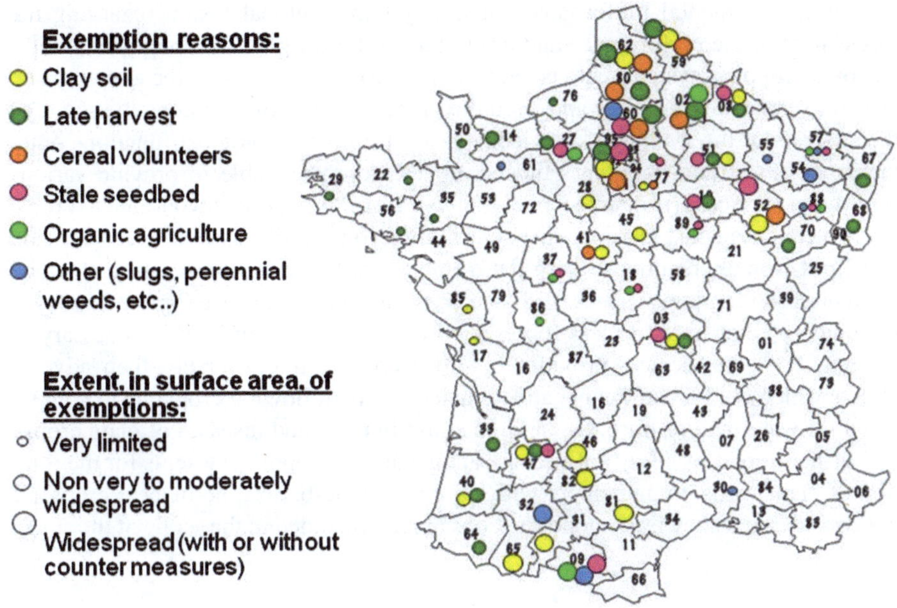

Fig. 1.4 Location and extent (in surface area) of the various reasons motivating exemptions from planting cover crops by French *département* (Source: prefectural and regional orders, as well as orders at *département* level)

unteers, (iv) the need to create stale seedbeds, particularly in organic farming, and, finally (v) the negative impact of cover crops on biotic factors (Fig. 1.4).

1.2 Study Objectives and Approach

1.2.1 Study Objectives

In their commissioning letter, already mentioned, the French Ministries for Agriculture and Ecology tasked INRA with conducting "a review of the available knowledge on nitrogen management during fallow periods, specifying the knowledge and research, as well as the uncertainties and questions requiring further investigation, with a view to extending the use of autumn cover crops". This study therefore attempts to analyse, as a priority, practices mentioned in the regulation policies, i.e. catch crops, oilseed rape and cereal volunteers, finely chopped straw buried or left on the surface. Furthermore, in addition to its role in nitrogen dynamics during the autumn period, fallow period management can also have other effects that are worth examining, with a broader view to moving towards more environmentally-friendly cropping systems. Since the implementation of a practice from which a benefit is expected for a given objective must not lead to a poorer overall environmental performance of the system, a global vision regarding the assessment of the performance and limitations of fallow period management methods on a cropping system scale is necessary. Bearing this in mind, the expert group tasked to carry out this study approved the initial proposal of extending the scope of the analysis of the available knowledge to a set of agro-ecological functions other than that of "nitrate-trapping". Indeed, cover crops are liable to provide various ecosystem services: (i) maintenance of physical soil properties favourable to crops, (ii) protection against erosion, (iii) contribution to pest, disease and weed control, (iv), limitation of greenhouse gas (GHG) emissions. These functions can be obtained by major crop families, such as Brassicaceae (mustards), Poaceae (grasses) or leguminous crops, or by other less represented crops. Consequently, it is necessary to evaluate both the services provided by cover crops and, conversely, their adverse effects which could result in negative impacts on the following crop, due to competition for water and/or nitrogen, or an increase in pests and diseases of main crops.

Furthermore, this study only covers annual arable cropping systems for the rotation of main crops, in agreement with the contracting bodies; permanent crops are not considered. Likewise, secondary crops are not included in the scope of the study.

1.2.2 Study Approach and Organization

The "studies" conducted by the Delegation of Scientific Expertise, Foresight and Advanced Studies (DEPE) of the INRA are exercises derived from Collective Scientific Expertise (ESCo) studies developed at the INRA since 2002 and for which specific standards are being developed. These ESCo studies are generally carried out at the request of public commissioning bodies – the Ministries for Agriculture and the Environment, notably – under the responsibility of INRA, by a multidisciplinary group of scientific experts, in charge of collectively analysing the international scientific literature in order to identify information useful for shedding light on issues raised by public managers. But the scientific literature alone is not always enough to provide useful responses to aid public decision-making. The DEPE thus proposed an approach that includes an ESCo-type scientific literature analysis component, supplemented, depending on the cases, by an analysis of technical publications, an original data processing approach, simulations or field surveys. In the present case, the study is based on simulations.

In this study, the expert group[3] is composed of researchers from the following research bodies or higher education establishments: INRA, IRSTEA (*Institut national de recherche en sciences et technologies pour l'environnement et l'agriculture* – French National Research Institute of Science and Technology for the Environment and Agriculture), Agroscope (Switzerland), ENSAT (*Ecole nationale Supérieur Agronomique de Toulouse* – National School of Agronomics, Toulouse) and the University of Toulouse. The results of the analysis of the litera-

[3] Experts are listed on the inside back cover.

ture component are presented in Sect. 2 of this document and the additional component, made up of simulation studies, is presented in Sect. 3. The texts for the full report have been reviewed by experts external to the study.

Analysis of the Literature

This component follows the principles of ESCo studies, i.e. sets out an analysis of the international scientific literature (articles published in peer-reviewed journals, as a priority, contributions to books published by recognized scientific publishers). The experts selected studies from all geographic sources for certain aspects of the analysis (description of generic mechanisms, etc.), but restricted their analysis to zones presenting comparable conditions (soil and climate, agricultural) to those of the French territory with respect to the local consequences of these mechanisms. As a result, studies conducted in Europe and North America were given priority. This bibliographic component encompasses all the functions that cover crops may fulfil. The technical literature is included insofar as the sources on which it is based (published experimental data, conditions for obtaining the data clearly defined, etc.) are specified.

Ultimately, the study drew on 795 bibliographic references, including 768 scientific publications and 27 technical documents. Over half of these references date from after 2005. They can be broken down as follows: (i) the analysis of cover crop management techniques (Chap. 3 of the report) was based on 74 references, (ii) nitrate and water during fallow periods (Chap. 4) was based on 256 references, (iii) the abiotic effects of cover crops on erosion, the physical properties of the soil, carbon sequestration and the greenhouse gas balance (Chap. 5) drew on 21 references, and (iv) the biotic effects of cover crops on weeds, microflora and fauna (Chap. 6) drew on 344 references.

Simulations

The conclusion that the data available in the literature will not provide the references necessary for every soil, climate and agronomic situation concerned by the planting of cover crops, prompted the decision to use simulations. This simulation work primarily aims to assess the impact of cover crops on nitrate leaching, water balance and the next crop, depending on the crops used, their management and in comparison with alternative practices, in a broad range of soil and climate conditions and cropping systems, representing the diversity of conditions found in mainland France. Two additional questions were also tackled through the simulation: the success of cover crop emergence and the number of days suitable for their mechanical destruction.

Following the first meeting of the study monitoring committee,[4] a technical group composed of engineers responsible for supplying data, providing information and discussing the simulation work was set up as an additional source of expertise concerning this simulation work. Concretely, this group is composed of engineers from the technical institutes most directly concerned (Arvalis, CETIOM (*Centre technique interprofessionnel des oléagineux et du chanvre* – Technical inter-professional centre for oleaginous crops and hemp), ITB (*Institut Technique de la betterave* – Technical Institute for sugar beet, ITAB (*Institut Technique de l'Agriculture Biologique* – Technical Institute for Organic Agriculture)) and Chambers of agriculture, competent in the field of fallow period nitrogen management and an engineer from a Belgian development body.[5] It met three times during the study. It was consulted regarding the relevance of the simulation plan and asked to supply experimental data. It was also asked to check the consistency of simulation results (in terms of the directional trends of effects, the magnitude and hierarchy of effects on the reduction of nitrate leaching and the impact on the next crop), in view of the expert knowledge of the engineers or references resulting from *in situ* trials, at testing stations or on farms.

[4]The study monitoring committee is made up of representatives from ACTA (network of animal and plant industry institutes), APCA (*Assemblée Permanente des Chambres d'Agriculture* – Permanent Assembly of Chambers of Agriculture), Coop de France, ONCFS (*Office National de la Chasse et de la Faune Sauvage* – National Office for Hunting and Wildlife) ONEMA (*Office National de l'Eau et des Milieux Aquatiques* – National Office for Water and Aquatic Environments).

[5]Jean-Pierre Cohan (Arvalis – *Institut du végétal* (Plant Institute)), Rémy Duval (ITB), Tanegmart Redjala-Ounnas (CETIOM – CETIOM), Sébastien Minette (Poitou-Charentes regional chamber of agriculture), Alain Lecat (Nord Pas-de-Calais regional chamber of agriculture and ITAB), Nathael Leclech (Lorraine regional chamber of agriculture), Damien Ronget (Côte d'Or regional chamber of agriculture), Frédérique Hupin (Nitrawal, Belgium).

Chapter 2
Main Lessons Drawn from the Analysis of the Literature

Eric Justes, Nicolas Beaudoin, Patrick Bertuzzi, Raphaël Charles, Julie Constantin, Carolyne Dürr, Alexandre Joannon, Christine Le Bas, Bruno Mary, Françoise Montfort, Laurent Ruiz, Jean-Pierre Sarthou, Véronique Souchère and Julien Tournebize

2.1 Cover Crop Management Techniques

Planting and destruction are the two main phases in cover crop management techniques. Their dates and implementation conditions play a key role in terms of the various functions targeted for these crops (nitrate-trapping, erosion prevention, weed control, effect on next crop, etc.). The analysis of the literature therefore focused on cover crop planting and destruction techniques. The aim was to identify the different methods and, if possible, their effectiveness with respect to the success

---

The original version of this chapter was revised.
An erratum to this chapter can be found at https://doi.org/10.1007/978-94-024-0986-4_5

E. Justes (✉)
UMR AGIR (AGroécologie, Innovations & teRritoires) 1248, University of Toulouse, INRA, INPT, Centre de Recherches INRA Occitanie-Toulouse,
24 Chemin de Borde Rouge, Auzeville, CS 52627, Castanet Tolosan 31326, France

UMR SYSTEM Univ Montpellier, Cirad, Inra, Montpellier SupAgro, CIHEAM-IAMM,
2, place Pierre Viala, Bat. 27, Montpellier 34060 Montpellier cedex 2, France
e-mail: eric.justes@inra.fr; eric.justes@cirad.fr

N. Beaudoin
UPR 1158 AgroImpact de Laon-Mons, Site de Laon, Pôle du Griffon,
180, rue Pierre-Gilles de Gennes, Barenton-Bugny 02000, France
e-mail: nicolas.beaudoin@inra.fr

P. Bertuzzi
INRA, US1116 AgroClim, Domaine St Paul - Site Agroparc, 228 route de l'Aérodrome,
84914 AVIGNON Cedex 9, Avignon CS 40509, France
e-mail: patrick.bertuzzi@inra.fr

R. Charles
Département Vulgarisation, Formation, Communication, Institut de recherche de l'agriculture biologique FiBL, Av. des Jordils 3, CP 1080, Lausanne 1001, Switzerland
e-mail: raphael.charles@fibl.org

E. Justes (ed.), *Cover Crops for Sustainable Farming*,
DOI 10.1007/978-94-024-0986-4_2

of sowing and destruction. The issue of the implementation of these techniques by farmers was then examined by analysing the constraints in terms of organization of labour and implementation costs.

The scientific literature proved to be sparse: fallow period management techniques as such are the subject of little scientific research (apart from in the methodological developments reported to complete certain trials), whereas they are the

J. Constantin
UMR AGIR (AGroécologie, Innovations & teRritoires) 1248, University of Toulouse, INRA, INPT, Centre de Recherches INRA Occitanie-Toulouse, 24 Chemin de Borde Rouge, Auzeville, 31326 Castanet Tolosan cedex, Toulouse CS 52627, France
e-mail: julie.constantin@inra.fr

C. Dürr
INRA, UMR 1345 IRHS, 42 rue George Morel, Beaucouzé 49071, France
e-mail: carolyne.durr@inra.fr

A. Joannon
UMR 0980 BAGAP (Biodiversité Agroécologie et Aménagement du Paysage), Université Bretagne Loire, INRA, Agrocampus Ouest-ESA, Rennes 35042, France
e-mail: alexandre.joannon@inra.fr

C. Le Bas
INRA, US InfoSol, 2163 Avenue de la pomme de pin, Ardon, CS 40001, Orléans cedex 2 45075, France
e-mail: christine.le-bas@inra.fr

B. Mary
INRA, UPR 1158 AgroImpact de Laon-Mons, Site de Laon, Pôle du Griffon, 180, rue Pierre-Gilles de Gennes, Barenton-Bugny 02000, France
e-mail: bruno.mary@inra.fr

F. Montfort
INRA, UMR IGEPP, Rennes F-35042, France

L. Ruiz
UMR SAS 1069, Université Bretagne Loire, INRA, Agrocampus Ouest-ESA, 65 Rue de Saint Brieuc, Rennes 35042, France
e-mail: laurent.ruiz@inra.fr

J.-P. Sarthou
UMR AGIR (AGroécologie, Innovations & teRritoires) University of Toulouse, INRA, INPT, Centre de Recherches INRA Occitanie-Toulouse, 24 Chemin de Borde Rouge, Auzeville, 31326 Castanet Tolosan cedex, Toulouse CS 52627, France
e-mail: jean-pierre.sarthou@inra.fr

V. Souchère
UMR SAD-APT INRA, AgroParisTech, Université Paris-Saclay, Avenue Lucien Bretignières, BP 1, Thiverval-Grignon 78850, France
e-mail: veronique.souchere@inra.fr

J. Tournebize
IRSTEA - Centre d'Antony, UR Hydrosystèmes et Bioprocédés / Hydrosystems and Bioprocesses Research Unit, Equipe ARTEMHYS, 1 rue Pierre-Gilles de Gennes CS 10030, 92761 ANTONY cedex, France
e-mail: julien.tournebize@irstea.fr

focus of attention by technical institutes and chambers of agriculture in particular, and technical innovation in this field is considerable. In the context of this study, publications concerning agricultural equipment, machinery and technical and mechanical innovations were not retained; for more details on the range of technical solutions currently being developed, readers may refer to the publications of these organizations.

The two simulation studies concerning the sowing-emergence phase[1] and evaluation of the days available for destruction of cover crops in the autumn,[2] are presented in Sect. 2.3 of the document.

2.1.1 Cover Crop Planting Techniques

The aims of the analysis of the literature were to identify and describe the different cover crop planting methods, to evaluate the research conducted on them and to identify the range of planting methods to be tested in cover crop emergence simulations.

The Methods Examined in Scientific Publications

Few bibliographic references specifically concern the study of technical planting variants for cover crops. In the majority of publications, these methods are mentioned since they contribute to experimental description of the situation studied, but they are not the central focus of the research.

These studies are often driven by a requirement to control erosion problems. The possible effects of these problems on the next crop are analysed. The results are sometimes contradictory in terms of the effects of the presence of the cover crop on soil moisture and temperature at the time of sowing the next crop and the reasons for these differences in effects are difficult to determine. In the studies identified, planting techniques (soil tillage, seed density, sowing date) are experimental variants for sowing a cover crop, for which the impact on speed of soil covering, the supply of nitrogen in the case of a leguminous crop, the reduction of weeds, soil structure, biodiversity, etc. are evaluated, as is the yield of the next crop.

Only one study,[3] conducted in France, describes in detail the variations in results obtained in terms of cover crop planting (emergence rates and speeds, reasons for non-emergence, physical conditions at the time of sowing). It indicates that the variations in the time between sowing and emergence are broader (a few days to a few weeks) than those for median emergence rates, which are always relatively high (>85 %) irrespective of the place, sowing date and techniques used.

---

[1] See report, Chap. 8.

[2] See report, Chap. 9.

[3] See report, Chap. 3 and Dorsainvil et al. (2005).

Ultimately, this analysis of the literature makes it possible to describe different technical variants for cover crop sowing, but only very rarely does it provide results in terms of the condition of the seedbed at the time of planting and control of the result in terms of uniformity and density of the cover obtained.

Variants Described in Technical Documents

Most of the references describing different practices, their effects and various attempts to innovate in terms of cover crop planting are produced by technical institutes, which conduct experimental and demonstration studies on these practices.

For planting, various types of surface stubble ploughing equipment may be used; depending on the equipment and its setting, the tilling depth varies. In contrast, ploughing prior to sowing of the cover crop is not mentioned.. Different sowing methods are also used, ranging from sowing using fertilizer spreaders (centrifugal or boom), to the use of tined or disk DS (direct seed) seed-drills, combined or not with a driven tool.

The effects of different sowing dates are presented, along with the latest dates not to be exceeded for broad French climatic regions in order to obtain sufficient cover (particularly for leguminous crops, but also for other crops), or, conversely, the earliest date to be complied with in order to avoid flowering (in the case of mustards) and a cover crop that will be too difficult to destroy.

The emergence rates obtained for these various trials can be relatively variable between years and depending on the soil tillage and sowing methods used. Crop and variety choices are discussed on the basis of regions and sowing periods. However, apart from the effects of the seed size on control of grain distribution (at the surface and in terms of depth), any potential effects of the germination and health characteristics of seeds are not indicated.

Regarding oilseed rape or cereal volunteers, the technical documentation indicates that the amount of grain lost during harvesting, if correctly scattered, generally makes it possible to obtain adequate plant cover, as long as emergence is successful. The latter can be delayed by a few weeks in cereals due to dormancy phenomena, but this process is not documented.

Consequences for the Emergence Simulation Study

The study of the literature makes it possible to describe the various planting methods, but does not give any precise indication of the variability of emergence results or the assessment of the risks of seed failure. This prompted the simulation study, using the SIMPLE model, of the variability of cover crop emergence depending on soil and climate conditions and sowing methods (frequency analysis over 20 years).

2.1.2 Cover Crop Destruction Techniques

Cover crop destruction methods are not the focus of any specific studies; they are discussed when they appear to be an important factor for the effectiveness of cover crops. The cover crop destruction date plays a key role in terms of the expected functions. It must also take into account the requirements of the next crop in terms of physical soil condition, availability of nutritional components (nitrogen, in particular) and the presence of inhibitory substances (allelopathy). Destruction is based on the use of mechanical or chemical methods, or on natural mechanisms such as senescence or frost.

Currently, glyphosate is the main substance used for chemical destruction, possibly combined with 2.4-D. The amounts of herbicide required vary significantly depending on the cover crops to be destroyed and the weeds present. For example, for some cover crops, such as ryegrass, destruction may require high doses of glyphosate, without achieving total control, or require several applications to be effective. The repercussions of the systematic and widespread use of this substance have been extensively documented in recent synoptic articles. However, the impact of introducing cover crops into cropping systems on the frequency of glyphosate use has been little documented.[4]

Ploughing represents a traditional method for the mechanical destruction of cover crops since it combines cover crop destruction with the usual autumn soil tillage for clay-loam and clay soils. Chopping or the use of a crimper-roller are other methods employed for the mechanical destruction of the aerial parts of crops and represent an alternative to chemical destruction using glyphosate. The use of a crimper-roller is a potentially effective method of creating a layer of mulch residue on the soil surface. However, the effectiveness of this mechanical operation is variable and depends on the crop, the amount of biomass cover and the weather conditions at the time of the operation. The technical literature makes extensive reference to these different methods.

In regions prone to such conditions, natural destruction by frost is also possible. The choice of crop (frost-sensitive crops) is the key when it comes to taking advantage of this "climatic potential". Frost value data are extensively available in the technical literature, but the information sources are not always indicated, which sometimes raises questions with respect to their reliability.

In conclusion, the management techniques most frequently described in the scientific articles consulted often involve the destruction of cover crops using mechanical methods or the application of a herbicide before the next crop is planted. However, the literature raises the issue of a number of risks related to the routine use of glyphosate (induction of resistance in weeds, transfer of the substance or its secondary metabolite (AMPA), and pollution of aquatic ecosystems).

---

[4] The only reference available concerns Switzerland where, between 1992 and 2004, the number of glyphosate applications in arable cropping systems increased from fewer than 0.05 applications per field and per year to around 0.15, related to the simplification of soil tillage and the systematic introduction of cover crops.

2.1.3 Constraints in Terms of Organization of Labour and Economic Constraints on a Field Scale

Despite the relevance of examining whether the organization of labour in farms is compatible with cover crop management, this question has not been tackled by the scientific community. No studies have been conducted concerning the feasibility of this practice on a farm scale in different climate, soil and farm type situations. Partial studies relative to the time required for the work and the days available, as well as the costs of the management techniques, do however exist. These demonstrate a high level of variability, in terms of the time required for the work, the cost of the different types of management techniques, and the days available for cover crop destruction. This highlights the need for a local approach on which to base fallow period management choices, taking into account the soil and climate characteristics, as well as the characteristics of the farm in terms of cropping systems and available equipment and labour.

An analysis on a field scale of labour times and the additional costs incurred by planting a cover crop was nonetheless conducted on the basis of references from the different technical documents. The results were expressed in terms of costs and additional labour time per management technique, in comparison with a situation in which the ground was left bare. It is known that fallow period management using volunteers generates less labour time and costs less. Regarding methods using catch crops, the additional cost, excluding labour, including sowing and destruction, ranged from between €30 and €92 per hectare, to which must be added €14–€60 per hectare in seed costs. In addition, these methods incur 0.6–2.2 additional hours of work per hectare. If this work is not done by the farmer, it is then necessary to count paid labour at a cost of between €9 and €33 per hectare. However, it should be noted that the majority of management methods involve less than 1.5 h/ha, i.e. an additional labour cost of less than €22.5/ha.

As far as clay soils are concerned, the difficulty of managing cover crops may reside in the soil preparation traditionally carried out rather than in the management of the cover crop itself. These types of soil are generally ploughed at the end of the summer in order to subject them to alternating wetting and drying, or in the autumn so that they can then benefit from alternating freezing and thawing. Ploughing too late in the autumn as a result of a delay due to the late destruction of cover crops, or even in the spring, can be problematic due to the soil moisture conditions: if the soil cannot be tilled or the result is unsatisfactory (compaction, etc.), the yield of the next crop can be compromised.

The lack of information in the literature prompted the implementation of a simulation study to evaluate the days available for destruction of cover crops, depending on the soil surface texture (see Sect. 3.2.3). This study only takes into account the biotechnical component of the days available (soil and climate conditions and factors related to the management technique), excluding socio-economic factors. However, it makes it possible to estimate, for a broad range of soil and climate conditions, the days potentially available as a function of the destruction date.[5]

[5] See report, Chap. 9.

2.2 Nitrate and Water During Fallow Periods

A wealth of international documentary resources relative to the "nitrate-trapping" (catch crops) and "green manure" effects of cover crops is available. These resources examine a broad variety of issues but they generally focus on local application and often lack a generic approach in terms of process analysis. This makes it difficult to extrapolate the results obtained at a specific site to other soil and climate situations. Analysis of the articles deemed to be the most relevant made it possible to reach positive and quantitative conclusions with respect to the effectiveness of catch crops. It also highlighted the factors that determine the effectiveness of catch crops, as well as the directional trends of their effects. However, the relationships were not determined and no quantification or modelling was performed, making it difficult to extrapolate the results. Consequently, this analysis of the literature demonstrated the need to perform simulations in order to obtain quantitative answers insofar as the scope of definition of the model used allows this.

2.2.1 Soil Nitrogen, the Origin of Nitrate and Leaching

Soil Mineral Nitrogen

The total nitrogen pool in the soil represents very significant quantities on a field scale (Fig. 2.1); the great majority of this nitrogen (over 98 %) is in organic form. Mineral nitrogen quantities in the soil – a very small percentage of the total – range

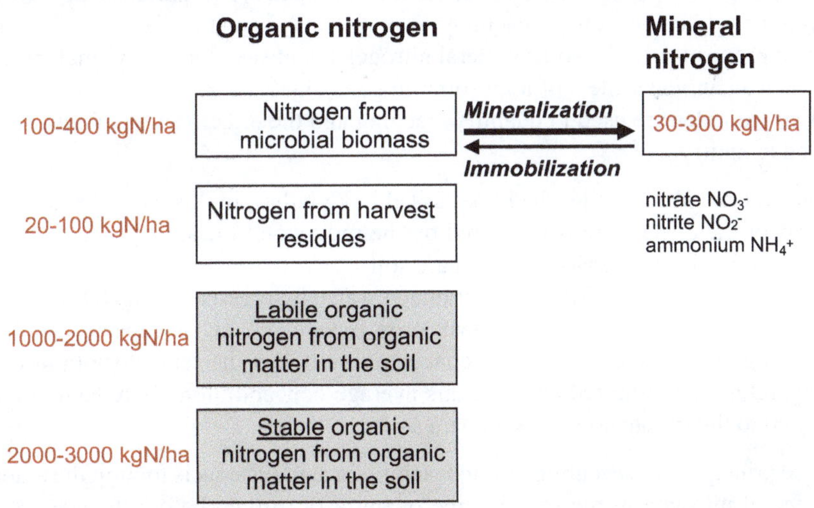

Fig. 2.1 Nitrogen compartments in the soil (as per Corpen, 1993) The values in *red* indicate the amounts of N/ha in the soil

from 30 to 300 kg/ha, whereas arable soil with an organic matter content of around 1.5 % contains about 3–5 tonnes of total nitrogen per hectare.

Many flux exchanges and transformations occur in the soil over the course of the season. The complexity of the nitrogen cycle is characterized by constant exchanges between organic and mineral forms.

Nitrate in Soils

The amount of nitrate in the soil varies over time and is the result of several antagonistic phenomena: addition by the application of fertilizer, mineralization of organic matter in humus and residues, atmospheric nitrogen deposits, gaseous losses (volatilization and denitrification), immobilization of nitrogen during decomposition of residues (nitrate uptake by microorganisms leading to its immobilization in the form of organic matter), absorption by plant cover and symbiotic nitrogen fixation in leguminous crops. These phenomena are primarily dependent on environmental conditions (temperature, moisture content, oxygen content of air in the soil, pH) and the availability of labile carbon (residues).

Nitrate Leaching and the Principles of Its Prevention

Leaching consists of the migration of solutes, including nitrate, with drainage water, beyond the reach of plant roots. It is a biophysical process and its intensity is determined by the interaction between the cropping system, farming practices and soil and climate factors; this intensity is therefore site-specific and, more particularly, year-specific, due to its reliance on rainfall amounts. The amounts of nitrogen exported by drainage waters (leaching) as nitrate may sometimes account for a significant proportion of the soil's mineral nitrogen but always for a very small proportion – on an annual scale – of total soil nitrogen (Fig. 2.1).

Two variables are used to quantify leaching and the effectiveness of cover crops for reducing it:

- the amount of nitrate leached (expressed in kg N/ha); in this case, the effectiveness of the cover crop is measured by the reduction in leaching (i.e. the nitrate loss avoided) in comparison with bare soil;
- the nitrate concentration of the drainage water (expressed in $mgNO_3/l$); in this case, the effectiveness of the cover crop is assessed by its capacity to obtain drainage water with an average concentration below the regulatory limit of 50 $mgNO_3/l$, or by the reduction in this average concentration (% reduction compared to the nitrate content with bare soil).

The principle of preventing nitrate leaching in agrosystems is to limit the amount of mineral nitrogen in the soil, i.e. the quantity of nitrate, before the start of the drainage period, using tactical methods (without affecting crop rotation) or, if necessary, strategic ones (change of cropping system and, in particular, of the crop rotation). Sowing a cover crop is one of the tactical options for trapping nitrate during the summer and autumn seasons without making any profound change to the cropping system. It may be necessary to maintain cover crops over the winter in mild and very rainy oceanic climates in order to effectively reduce leaching.

The nitrogen balance of the main crop is a primary factor determining the mineral nitrogen pool present at the start of the fallow period. The judicious use of fertilizer application minimizes the residual soil mineral nitrogen content (nitrate and ammonia content) at the time of harvest. However, as significant, uncertainties are associated with the estimation of the optimal fertilizer dose, fertilizer is often found to have been applied in excess. Above the optimum fertilization level, the residual amount of nitrogen present at the time of harvest increases in close correlation with the excess nitrogen fertilizer applied. So, for example, in the case of moderate over-application of fertilizer, for every 2 kg of excess nitrogen applied, an additional 1 kg of residual nitrogen (in mineral form) is found, on average, at the time of harvest.

After harvesting, the mineral nitrogen pool increases due to the mineralization of organic matter in the soil. This leads to the release of ammonium ions which, in turn, adds to the nitrate pool as a result of nitrification. In temperate climates, the two processes of mineralization and nitrification are closely coupled (nitrification occurs almost simultaneously with the release of ammonium ions produced as a result of mineralization), which explains the dominant presence of nitrate in the soil. The mineralization rate depends on the soil type and climatic and farming conditions. The amount mineralized is therefore highly variable, ranging from around 40–70 kgN/ha in the autumn in temperate conditions, depending on organic nitrogen content, temperature and soil moisture. Mineralization occurring in summer and autumn determines the risks of leaching to a greater extent than the residual mineral nitrogen amount present at the time of harvest i) in situations in which the fertilizer dose has been properly determined, and/or ii) in cases in which the drainage water is enough to completely renew the soil water content corresponding to the soil water holding capacity.

The management of harvest residues also affects the risk of leaching, depending on their mass and nitrogen content (or carbon (C) to nitrogen (N) ratio, C/N). The incorporation of harvest residues generally leads to net mineral nitrogen immobilization in the soil. However, although the potential for mineral nitrogen immobilization in the soil is very high in when finely chopped residues are evenly mixed in the soil (as indicated by incubation trials), the amounts immobilized in the field are generally only around 10–30 kgN/ha after 6 months of decomposition. The decomposition of mature residues from the previous crop temporarily mobilizes the mineral nitrogen in the soil (adding to the readily mineralizable pool of organic nitrogen in the soil). The amount of mineral nitrogen mobilized in this way depends on the extent of contact between the residues and the soil and, therefore, the type of soil tillage and incorporation depth. Net mineral nitrogen immobilization only affects the layer where the residues are incorporated, explaining their low impact in terms of reducing leaching.

2.2.2 *Nitrate Trapping Using Catch Crops*

On a yearly scale, the introduction of a cover crop leads, in the very great majority of trials, to a significant reduction in nitrate losses due to leaching. But the intensity of this varies depending on the site, the crop and the planting and destruction conditions. This variability can be explained by the local nature – in terms of time and space – of both cover crop growth and leaching conditions. The amount of nitrate trapped by root absorption, and which is contained in the biomass of a catch crop, can potentially reach 100–150 kgN/ha with Poaceae and over 200 kgN/ha for Brassicaceae. The amount of nitrogen in leguminous crops may also be high, but this is partly due to symbiotic nitrogen fixation, meaning that the amount of nitrate trapped is lower for this family of crops. The quantity of nitrate uptake depends on the date, planting conditions, growth time and type of cover crop, as well as the amount of mineral nitrogen available in the soil. The amount of nitrogen uptake and the variation in soil mineral nitrogen between cover crop and bare soil treatments in the autumn are indicators of the potential reduction in leaching, except in the case of leguminous crops. The actual reduction in leaching depends on the timely competition between the processes of nitrate transfer deep in the soil and of uptake by the cover crop, and then on the rate of residue mineralization following their burial.

The Dynamics of Nitrogen Uptake by the Cover Crop: Conditions Determining Effectiveness and Specific Case of Late Autumn Harvests

Soil and climate characteristics have a significant influence on the rate of nitrate transfer, but also on the rooting depth of the cover crop and the mineralization of residues. The risk of leaching is greater in rainy climates and in shallow, rough-textured soils. The absorption period of the cover crop must be adapted to the drainage dynamics. It is therefore essential that the cover crop traps soil mineral nitrogen before drainage becomes intense and leaching occurs. This constraint leads to the existence of an optimal emergence date, by crop type and by region, but also a latest sowing date beyond which the catch crop becomes ineffective. This raises the question of fallow period management following harvesting of the main crop in the autumn; residue management has only a small limiting effect on leaching and the growth of catch crops sown at the end of October is often reduced by low temperatures and global radiation. The practice of undersowing a cover crop with a main crop harvested in the autumn may give good results in Southern European climates (as observed in trials in Italy), but it needs to be the subject of specific studies, taking into account the effects of competition with the main crop in order to increase the success of this method. Furthermore, both weed control and fertilizer management have to be considered in order to effectively control winter nitrate losses.

The Dynamics of Nitrogen Uptake by the Cover Crop: Case of Clay Soils

In the case of clay soils, planting and destruction constraints require the specific adjustment of cover crop management, but the review of the literature was unable to provide a comprehensive analysis of this question. Simulations with different sowing and early destruction dates are useful when it comes to examining this specific issue more closely.

Reducing the Nitrate Concentration of Drainage Waters

The effectiveness of cover crops in terms of reducing the nitrate concentration (or content) of drainage (or percolation) water is significant in 90 % of the studies analysed, but with a very high level of year-to-year variability and depending on soil and climate conditions. This reduction is observed in almost all situations, in comparison with a bare soil control. It is, to a very large extent, due to trapping by the cover crop of mineral nitrogen in the soil profile, leading to a reduction in leaching, not a reduction in the drainage volume. Although the effectiveness of cover crops depends on the drainage level, their impact on drainage or on the available water capacity present in soils at the end of winter is low. The average reduction in drainage under the soil profile is around 30–50 mm, i.e. 6–10 % of the annual volume, with reductions of up to 22 % in the trials reported in the literature.

Choice of Type of Cover Crop

The crop type is a determining factor in the management of timely competition and the return of nitrogen to the next crop. However, this is not the only factor that determines effectiveness in that it is first of all necessary for cover to be in place at the end of the summer and during the autumn to absorb mineral nitrogen from the soil. Then, depending on the date on which leaching starts, a crop type with the capacity to quickly absorb mineral nitrogen deep in the soil will make it possible to obtain the effect expected from timely competition, depending on soil and climate conditions. Generally speaking, the reduction in amount of leached nitrate and in nitrate concentration of drainage waters is, on average, twice as high for non-leguminous cover crops as it is for leguminous ones. However the effectiveness of leguminous crops is usually significant, despite being markedly lower than that of non-leguminous crops. Given their capacity to fix atmospheric nitrogen and make a greater proportion of this available to the next crop due to their higher nitrogen content, mixtures of leguminous and non-leguminous cover crops appear to combine the advantages of "nitrate-trapping" catch crops and "green manures". However, the references – less numerous for mixed crops – demonstrate that the reduction in leaching is not always as high as with non-leguminous cover crops. This review of the literature, which doubtless needs to be supplemented by specific studies examining mixed crops, makes it possible to draw up an initial synopsis of the magnitudes of these effects on the basis of the crop types used as cover crops (Table 2.1).

Table 2.1 Summary of the effects, advantages and limitations of different cover crop types

Factors effects	Non-leguminous poaceae	Non-leguminous brassicaceae	Leguminous	Mixtures legum. & Non-legum.
Requirements	Relatively early sowing	Early sowing	Very early sowing	Adjust the mixture
	Not or not very frost-sensitive	Frost-sensitive depending on type and temperature	Frost-sensitive	
Leaching reduction (rate/control)	30–80 %	30–90 %	0–40 %	20–60 %
				(to be confirmed[a])
Short-term nitrogen effect (N released as % N absorbed)	−20–+10 %	−10–+30 %	+10–+50 %	+10–+40 %
				(to be confirmed)
Advantages	Effectiveness with high nitrogen input levels or in oceanic climates	Broad effectiveness with high nitrogen input levels or in continental climates	Effectiveness with low nitrogen input levels	Intermediate effectiveness and plasticity/to N availability
				Less competition as undersown crop
Systems not recommended or to be avoided	Clay soil (/late destruction).	Clay soil if not frost-sensitive (/late destruction).	Systems with intensive nitrogen & effluent inputs	Nitrogen-intensive systems

[a] To be confirmed by a targeted study of the literature and simulations including a combination of crop types

Cover Crop Planting and Destruction Dates

The amount of nitrate trapped also varies depending on when the cover crop is planted and destroyed. In the North-Eastern quarter of France, for example, the amount of nitrogen trapped decreases significantly for crops sown after 10 September, or even 1 September for the "coldest" zones, and depending on the type of cover crop. Regarding destruction, although nitrate-trapping increases when the destruction date is delayed, there is a maximum date beyond which maintaining ground cover has no or little effect on trapping, although it does stop catch crop residues mineralizing too early. It is only once the cover crop has been destroyed that mineralization of its residues adds to the soil's mineral nitrogen pool. Field trials conducted in various soil and climate conditions on a rotation scale have demonstrated that the release of nitrogen following incorporation of a cover crop and its use by subsequent crops may be highly variable, ranging from net mineralization (positive effect on nitrogen availability) to, sometimes, net immobilization (negative effect on nitrogen availability).

2.2.3 *Effects Following Destruction of the Cover Crop*

Effects on the Next Crop

In addition to the amount of nitrogen required by the cover crop, the most robust indicator of the amount of nitrogen released and available for the next crop is the C/N ratio of the cover crop residues. Since the carbon content varies little (from 38 to 45 %), it is the nitrogen content (highly variable, from 1 to 4 % approximately) that determines the C/N value. This ratio is highly variable depending on the cover crop type and growing conditions, which are determined by the availability of mineral nitrogen in the soil (or the effectiveness of symbiotic nitrogen fixation for leguminous crops). So, for example, the C/N decreases when cover crops are planted in conditions of high nitrogen availability in the soil or for leguminous crops. The nitrogen input substitution effect ("green manure"), i.e. the reduction in fertilization of the next crop permitted by the incorporation of a cover crop, is between 50 and 60 % at the most, but, more usually, lies in the range of 10–40 %, because the C/N ratio of cover crops often ranges between 12 and 20.

Consequently, the use of pure or mixed leguminous crops, which induce low C/N values, often leads to an increase in the yield of the main crop that follows as a result of the release of nitrogen due to mineralization of cover crop residues. This effect is less conclusive when high levels of fertilizer have been applied to the main crop (increase effect less than proportional and situations that become non-limiting in terms of nitrogen content), whereas non-leguminous crops do not generally lead to an increase in yield in annual studies. Adequate fertilization should incorporate the potential depressive effects on mineral nitrogen levels caused by the cover crop or, conversely, potential positive effects on nitrogen return by mineralization of its residues. In order for these effects to be properly taken into account, it is essential to know the amount of nitrogen uptake by the catch crop and its C/N ratio at the time of destruction.

The Long-Term Effects

The repeated incorporation of cover crop residues can lead to cumulative effects in terms of the storage of organic matter in the soil and its subsequent mineralization potential. The latter increases in proportion to cumulative organic nitrogen inputs in the form of cover crops, reaching a plateau after 20–40 years, which depends on the type and C/N ratio of the cover crop residues. This additional mineralization generally increases the nitrogen uptake of main crops. Despite the increased mineralization triggered by the successive burial of cover crops, when these become an integral component of the cropping system and are routinely planted each time the crop rotation makes it possible, catch crops remain an effective way of reducing nitrate leaching in the long term. After several years (more than 10), however, it is necessary to reduce the amount of nitrogen fertilizer used on main crops to take into account this additional mineralization of the soil and hence maintain a high level of effectiveness of cover crops in terms of reducing leaching and nitrogen concentrations in drainage water.

The Effects on the Water Balance: Restoration of Soil Available Water Content and Effect on the Next Crop

The use of cover crops in temperate climates does not lead to any major changes in soil available water content in the springtime, or at the time of sowing the next crop, unless they are destroyed very late (for example just before sowing the next crop). Generally speaking, due to rainfall between destruction and sowing, the soil profile water content is restored to almost identical levels as for soil without a catch crop. Apart from in extreme cases, no significant effect has been found on the water supply for the next crop in the literature. This conclusion is consistent with that reached regarding the impact of cover crops on water drainage on a field scale, which is only moderately affected in temperate climates.

The impact of reducing drainage on the scale of a watershed or of the capture zone of an alluvial aquifer was not specifically examined in the bibliographic references. To do so, a hydrological study incorporating the spatial distribution of soils and crops, and water transfers under the soil profile should be carried out, on the basis of the hydro-geological scenario concerned by the analysis. However, this type of analysis did not fall within the scope of the present study.

2.2.4 Cover Crops Within the Cropping System

Complementarity Between Cover Crop Management Techniques and Multi-functionality

The main techniques of fallow period management in the autumn (for example, management of residues and sowing of a cover crop) appear to be potentially complementary in terms of reducing leaching since they do not apply to the same period nor to the same processes in the nitrogen cycle. In particular, the residue management type (incorporation into the soil versus mulching) has an effect on the dynamics of mineral nitrogen and water up to the depth of residue incorporation (a few centimetres with surface stubble ploughing), whereas a cover crop can absorb mineral nitrogen deep in the soil profile (to several tens of centimetres or, in some soils, more than 1 m). Residue management and planting cover crops can be coordinated in order to combine these effects, making it possible, for example, to limit the effects on drainage using a mulch made of the previous crop residues at the surface, reducing evaporation. More generally, this water and nitrogen management must be reconciled with the other fallow period objectives, such as harvest residue management, weed and pest control, soil protection and planting of the next crop. This indicates the need for further studies in order to optimize fallow period management on the basis of priority objectives.

Cover Crops and Redesign of Cropping Systems

The impact of the widespread use of cover crops on a regional scale depends on the structure of its cropping plan – in particular the proportion of fallow periods within the rotation, the calendar period of this fallow period, the date on which the previous crop is harvested and the next crop sown – without under-estimating the effectiveness of cover crops in certain crop rotations with a shorter fallow period. However, cover crops may also be a lever for redesigning cropping systems as part of a preventive approach to reduce leaching and a strategy aimed at achieving nitrogen autonomy for the cropping system, which may lead to a change in the cropping plan itself, depending on the objective sought. This issue of redesigning cropping systems is not tackled in the present study, but remains an important consideration in terms of the evolution of current cropping systems.

2.3 Effects of Cover Crops on Erosion, Physical Soil Properties and Carbon Balance

2.3.1 Effects of Cover Crops on Erosion and Physical Soil Properties

Overall, the analysis of the literature demonstrates the existence of a general consensus that the benefits of cover crops are not limited to their nitrate-trapping capacities (catch crops). Cover crops also help protect the soil surface against erosion and improve the structural status and physical properties of soil horizons reached by the roots. Consequently, the use of cover crops may be particularly beneficial in terms of:

- soil protection, especially in the winter, and facilitation of soil preparation for sowing the next crop. The improvement in the structural condition and physical properties of soil surface horizons is the result of mechanical surface-protecting effects of the cover crop against winter rain and soil aggregation during the development of the root system of the cover crop, the action of soil macrofauna on soil porosity and the resulting hydrodynamic properties, and modification of calendars for tillage, carried out in the autumn in the case of direct sowing of the next crop (without ploughing or very superficial tillage);
- improvement of the organic status of soils and the resulting physicochemical properties after burial of the cover crop.

Hence cover crops undeniably have positive effects in terms of reducing water-induced soil erosion. Their aerial biomass protects the soil surface from the impact of raindrops, while their root system – if it is dense and ramified – also helps to improve the soil's resistance to erosion. They promote the infiltration of water into the soil and at the same time slow down run-off by forming physical obstacles to water flows. However, the expression or occurrence of these positive erosion-reducing effects is not systematic. The effectiveness of cover crops depends on the quantities of aerial and root biomass produced, something that is linked not just to the type of crops, but also to how early they are sown, the conditions in which they are planted and the date of their destruction, related to the climate. However, not enough trials have been conducted in a temperate climate context to be able to rank the various crops according to the criteria analysed. The threshold – frequently cited in the literature and widely accepted – of 30 % soil cover required to obtain an effect against erosion would be worthy of consolidation through additional tests, particularly as a function of soil surface texture, slope and type of soil tillage.

Protective effects on the physical and hydrodynamic properties of soils and on compaction are primarily due to the structuring effect of root systems and an increase in organic matter in the soil. Over a fallow period, the effects of cover crops on porosity, structural soil stability and infiltrability at the soil surface are generally positive, but may be more or less marked depending on the crop type. However, in terms of intensity, these effects are secondary to those related to climatic variability, the type of soil tillage employed for planting the cover crop and, more generally, the cropping system within which the cover crop is integrated. Most publications concern trials carried out in conditions that differ from those in France, either in terms of climate, soils, cropping systems or cover crop types. Furthermore, many studies compare different cropping systems in which the cover crop is not the only variable analysed, making it difficult to determine the individual effect of the cover crop. For the processes studied, therefore, quantitative extrapolation of the results analysed to French systems is difficult and risky without conducting a more in-depth analysis of the study conditions and unfortunately these conditions are often inadequately detailed.

2.3.2 *Effect of Cover Crops on Carbon and Nitrogen Sequestration and Greenhouse Gas Emissions*

Although the primary benefit of cover crops – and the one that justifies their increasing use within European cropping systems – is the reduction in nitrate leaching, the repeated use of catch crops can also have an impact (positive or negative) on carbon sequestration in the soil and the emission of greenhouse gases (GHG), such as CO_2, N_2O and CH_4. It is widely assumed that: (i) cover crops have a low carbon-sequestering effect and (ii) they increase N_2O during the residue decomposition phase following crop destruction, (iii) ultimately producing a mixed result in terms of GHG balance. The analysis of the recent scientific literature largely contradicts these assumptions, indicating a broadly positive effect of catch crops on GHG balance. These results need to be confirmed by additional measurements through *ad hoc* trials.

The great majority of inventory studies carried out to determine variations in organic matter contents in soils, and hence carbon sequestration, demonstrate that cover crops induce the sequestration of carbon and organic nitrogen in soils. The values found in the literature indicate average sequestration of 300 kgC/ha, with a relatively broad variability (standard deviation of 150 kgC/ha). The sequestration intensity depends much more on the quantity of biomass produced and incorporated in the soil (aerial and root parts) than on the type of cover crop, although leguminous crops appear to promote sequestration, at least in tropical environments. A more thorough understanding of this is required to determine whether this is a "dose effect" (quantity of C incorporated) or whether it is related to the biochemical nature of the residues.

Finally, the rate of humification in the soil of carbon produced by cover crop residues (non-mature plants) appears to be comparable – or indeed higher – than that resulting from main crop residues (mature plants: straw), in contrast to what is generally assumed. This high rate of humification of catch crop residues explains the significant and important effect of carbon sequestration due to cover crops.

The studies conducted to measure the short-term impact of cover crops on N_2O emissions (only a few such studies have been conducted) show that non-leguminous catch crops can lead to a slight increase in GHG emissions following their incorporation. This slight increase in N_2O is 0.11 kg N/ha on average, but the variability between studies is very high (standard deviation of 1.12 kgN/ha). Nonetheless, the average N_2O emission level appears to be significantly lower than that calculated using current IPCC standards. No effect has been reported on CH_4 emissions. Moreover, for leguminous cover crops, the lack of references means that it is not possible to estimate their impact on N_2O emissions.

In temperate climates, the impact of non-leguminous cover crops on the GHG balance can be estimated by adding together their contribution to N_2O, emissions, carbon sequestration (which leads to a reduction in CO_2 concentrations in the atmosphere) and CO_2 emissions related to agricultural operations (sowing, destruction). N_2O emissions are converted into CO_2 equivalent by considering that the global warming power of N_2O is 296 times higher than that of CO_2. Generally speaking, cover crops have a positive effect on the GHG balance and this effect can be observed in every year they are planted. The average reduction is around −1 tonne of CO_2 equivalent/ha. The confidence interval of the GHG balance can be estimated using an upper limit corresponding to the most unfavourable case (high N_2O emissions and low carbon sequestration, which is unlikely) and a lower limit (low N_2O emissions and high carbon sequestration). This balance will be between +0.1 and −2.1 tonnes of CO_2 equivalent/ha for each year that a catch crop is planted as part of the rotation. It is therefore negative in the great majority of cases, indicating that the practice of planting cover crops is beneficial. However, it should be taken into account that carbon sequestration has been calculated over an average period of 15 years here, considering i) an asymptotic process for carbon sequestration, and ii) an annual GHG balance that is decreasingly negative over time due to carbon release as a result of the mineralization of organic matter in the soil. Nevertheless, this balance is likely to remain favourable in the long term.

Additional measurements therefore need to be made in order to more precisely quantify the impact of cover crops on greenhouse gas emissions and to evaluate their variability for the broad range of soil and climate conditions and cropping systems encountered in France.

2.4 Biotic Effects of Cover Crops on Weeds, Microflora and Fauna

The use of cover crops as nitrate-trapping catch crops or to protect the soil against erosion now needs to be considered in the light of sustainable development objectives in agriculture (reduced pesticide use) and hence their impacts on pests, diseases and weeds, functional biodiversity (organisms providing ecosystem services to agriculture: beneficial zoophagous organisms (natural enemies), pollinators, soil organisms, etc.) and cultural biodiversity (emblematic species, game). This section reports the now documented effects of cover crops on weeds, microflora and fauna of agricultural value (or hunting value for the latter group), be they pests or "useful" organisms.

2.4.1 Weeds and Fallow Period Management

The Mechanisms Involved

The objective of using cover crops to combat weeds is to replace weed flora that cannot be controlled with plant cover that can be controlled.

The effects of growing cover crops and their residues on weeds are based on various competitive and environment-modifying mechanisms (water, nitrogen, light, and other nutritional components). Cover crops as a living mulch compete directly with weeds for resources. The availability of these resources is also influenced by changes in the environment as a result of cover crops. This process concerns both the fallow period and the next crop. Allelopathy is (potentially) a different interaction phenomenon. It is chemical in nature and involves the emission by certain plants and/or their residues of substances that are toxic for other plants. These substances may have an impact on weed germination.

One of the difficulties encountered when analysing the effects of cover crops on weeds lies in the fact that the various mechanisms of action cannot be systematically differentiated from one another. Although the capacity of a cover crop to compete for abiotic resources is generally manifested by rapid growth and high biomass production, its effectiveness due to modification of the environment or via an allelopathic process is difficult to determine. It is assumed that the two mechanisms – competition and allelopathy – generally operate together, leading to a potentially high level of competition with weeds. Some studies focus primarily on allelopathy and the chemical substances responsible for this effect. However, the prospects for direct application regularly come up against the problem of the complexity of the processes involved.

Finally, it appears that only the Brassicaceae family demonstrates a sufficiently characteristic allelopathic activity to justify the explicit exploitation of this property. However, this property should not be seen as the only mechanism and the simple solution in terms of weed control using cover crops, and the factors determining the variation and management of allelopathic potential in the field are yet to be defined.

Effects of Cover Crops on Weeds

Planting a cover crop primarily modifies the chronological sequence of soil tillage and consequently constitutes a means of directly competing against weeds or making the environmental conditions unfavourable to their development. The range of crops that can be used as cover crops is extensive enough to address the diversity of soil, climate and agricultural conditions found in France. Choosing crops from different botanical families to those of main crops therefore makes it possible to diversify the cropping system and potentially break the growing cycle of certain types of weeds. The competitive effect of a cover crop on weeds depends on several factors, such as how quickly it becomes established, the growing duration and the persistence of biomass in live or dead form following destruction. The effects of a cover crop on weeds also depend on more or less prolonged modifications in soil fertility status and, in particular, lower availability of nutritional resources, with the use of leguminous crops representing a specific case in terms of nitrogen resources. Introducing a cover crop into the rotation also leads to changes in soil tillage, when the cover is sown and during its management before the next crop. The main challenges are to adapt the management of residues following destruction (incorporation into the soil or mulch on the soil surface) in order to maintain the weed control effect for as long as possible, or even to reduce the use of herbicide, at the end of the fallow period or at the start of the next crop.

It is clear that biennial and perennial species are more difficult to control using cover crops, due to their greater competitiveness as a result of their nutritional reserves (primarily located in the roots). These reserves generally mean that they demonstrate faster re-emergence and growth.

Cover Crops, Soil Tillage and Compatibility with the Stale Seedbed Technique

Any soil tillage has the effect of mechanically regulating weeds, but also constitutes a modification of the environment favourable to the expression of the seed stock present in the soil. Mixing up the soil has an influence on the development of weeds as a result of mechanisms such as exposure of seeds to light, which can end their dormancy, aeration of the soil and an increase in its temperature, the reduction of cover by residues, the elimination of inhibitory substances in the soil and an increase in soil/seed contact promoting germination.

The stale seedbed technique consists in preparing a seedbed that is fine enough to sow a crop with small seeds, to allow a proportion of the weed seed stock to germinate, and then to destroy the germinated seeds and seedlings that have emerged by tilling the soil or applying herbicides. This technique can be repeated several times before the main crop is sown, as is often the case in organic agriculture.

The combination of stale seedbeds and cover crops still needs to be studied therefore, in order to determine the conditions for their compatibility. It would appear that the management of annual weeds and perennial weeds with seed regeneration is compatible with the introduction of a catch crop into certain crop rotations, because the weeds that will have emerged at the same time as the cover crop will also be destroyed at the same time. In this case, the stale seedbed soil preparation phase does exist but it is followed by sowing of the cover crop, and the destruction phase is only performed much later, when the cover crop is destroyed. It must nonetheless be highlighted that planting a catch crop may pose problems for species in which the seeds are still too dormant when the cover crop is planted and hence may not germinate after soil tillage. The seed stock may also remain intact if soil tillage is carried out early after harvesting and post-harvest rain is insufficient to soak recently produced seeds that have fallen onto the ground. In these situations, weed seeds are not sensitive to the effect of stale seedbeds because they need to be soaked in order to germinate.

Recent studies (based on simulations using the FLORSYS model[6]) indicate that cover crops can reduce multi-species autumn weed flora thanks to soil tillage accompanied by the planting and destruction of catch crops. Hence, in systems

[6] Colbach et al. (2012). Using a sensitivity analysis of a weed dynamics model to develop sustainable cropping systems. II. Long-term effect of past crops and management techniques on weed infestation. Journal of Agricultural Science, Cambridge (in press). Colbach et al. (2012).

where soil tillage was initially carried out in springtime only, the introduction of catch crops reduces crop infestation with weeds, mainly because it requires soil tillage operations in the summer or autumn, playing the role of a stale seedbed. In systems in which a proportion of soil tillage was initially carried out in summer/autumn, the weed infestation of crops could, in contrast, increase if this soil tillage is not carried out when weed seeds are sensitive to stale seedbeds.

The case of perennial weeds demonstrating vegetative regeneration (rhizomes, runners), such as *Carduus* and *Cirsium* thistles in particular, can prove to be more problematic, although adaptations can be considered via the specific management of areas of the field (precision agriculture approach) presenting concentrations of perennial weeds. However, it is important to indicate that research needs to be carried out in this area, incorporating other perennial weeds with vegetative regeneration, such as *Rumex* (dockweed), *Convolvulus arvensis* (field bindweed) and of course *Elymus repens* (couch grass).

Studies Are Required to Quantify the Effects of Catch Crops on Weed Flora

The effects of introducing cover crops into the crop rotationas a result of the impact of plant cover (modification of the environment, competition and allelopathy) and changes in soil tillage practices may prove to be significant. It may even be necessary to rethink the weed control strategy: the use or otherwise of surface stubble ploughing to manage volunteers or weeds from the previous crop, the use of herbicide (in particular glyphosate) to destroy the cover crop and associated weeds, modification of the herbicides used and of their effectiveness for the next crop. However, the occurrence of new weed-related problems following the introduction of cover crops has not been demonstrated in the scientific literature. Cover crop management with a view to weed control may potentially lead to a reduction in herbicide applications but, to achieve this, the strategy needs to be adapted on the basis of the weed flora present and rationalized on the basis of the cropping system as a whole.

The need to take into account numerous factors determining the dynamics of weed flora and its specificity as a function of the cropping system and the type of soil tillage means that more research is required. The aim is to gain a better understanding of the effects of cover crops on weeds and hence to be able to accurately predict them. Studies of this type can now draw on modelling as well as on dedicated experiments.

2.4.2 Microflora and Fallow Period Management

Fallow periods can be managed in such a way as to break the epidemic cycle of pathogens present in the crop rotation practised, i.e. avoiding crop volunteers and cover crops acting as hosts for the main diseases. Soil tillage practices that help to speed up the breaking down of crop residues reduce the primary inoculum of pathogens. By reducing the projection of spores, soil cover may limit the spread of pathogens. However, all these potentially positive effects may be offset by effects that promote diseases under the influence of numerous determining factors.

The insertion of a cover crop into the rotation may actively disrupt the epidemic cycle of the pathogenic microflora in the soil, acting on different levels: via purely allelopathic mechanisms, but also by modifying the soil's receptiveness to diseases, through modifications induced by the cover crop and its subsequent burial and degradation in the soil. Alliaceae, Brassicaceae (mustards) and Poaceae (grasses) are all families with an allelopathic potential, the effects of which may be exerted by root exudation or after chopping of their tissues. Apart from avenacin (produced by oats), which has a specific action on Oomycetes, the other substances present numerous biological targets. The nature, quantities and persistence of the toxic compounds emitted are extremely variable depending on the agricultural, edaphic and biological parameters, as well as the techniques used during chopping and incorporation into the soil. Biofumigation (the agronomic application of the phenomenon of allelopathy) is based on the presence of various glucosinolates (GSLs) in the root exudates of Brassicaceae, which are subsequently transformed into isothiocyanates (ITCs). These are potentially toxic for soil-borne diseases. To maximize the effects, biofumigation is often carried out (particularly in legume crops) by growing a specific species in the fallow period and then chopping it into green manure and burying it, generally at the flowering stage. Some species of Brassicaceae are specifically selected because they contain high amounts of certain GSLs in order to produce effects on given pathogens. However, it should be noted that since these substances have a broad action spectrum, non-pathogenic microorganisms – or even beneficial ones – in the microflora may also be affected.

Apart from these allelopathic phenomena, simply planting a cover crop induces numerous physicochemical and biological changes in the soil. As a result it may sometimes have marked effects on the receptiveness of the soil to diseases, a phenomenon which is difficult to demonstrate and distinguish from other effects.

Regarding the burial of cover crop residues in the soil, the majority of studies highlight the changes induced in soil-borne microbial communities, i.e. stimulation of the activity of antagonistic populations indigenous in the soil, the stimulation being inversely correlated with the level of diseases. It seems that the effectiveness of a species, or even of a variety, is more marked when the natural antagonists are already well established in the soil (the concept of "soil health", which is controversial in the scientific literature). Phytopathogenic species with high saprophytic capacities are liable to multiply significantly and cause damage before natural biological regulation phenomena are manifested. Biofumigation is an interesting example to illustrate the complexity and multiplicity of the mechanisms at play when introducing cover crops during fallow periods: going far beyond a straightforward toxicity of secondary metabolites, such as ITC for Brassicaceae, complex microbiological interactions also explain the suppressive effects on pathogens.

Cover crops also have an impact on beneficial soil-borne microflora. Some species can have a stimulating effect on antagonistic fungi – very important natural enemies of nematodes – thereby amplifying natural regulation. Similarly, cover crops can help restore the diversity and density of mycorrhizal fungi, which play a major role in the uptake and transfer of nutrients, and can be disrupted, by monocropping for example. However, since Brassicaceae do not host mycorrhizal fungi, cover crops using plants from this family ultimately reduce pre-existing mycorrhizal populations by releasing ITCs which are toxic to soil fungi. Finally, generally speaking, the microbiological activity of soils depends greatly on the type of crop rotation practised and the use of cover crops, along with the addition of organic matter which can increase it, leading to a beneficial impact on several parameters influencing plant health, provided that few natural enemies in the microflora are destroyed.

2.4.3 Fauna (Invertebrates, Vertebrates) and Fallow Period Management

This section focuses on the effects of cover crops on fauna (micro-, meso- and macrofauna), but also on the epigeic and aerial fauna of the soil, be these crop pests, natural enemies or organisms that are neutral with respect to crops.

The first studies on phytoparasitic nematodes, carried out in the 1970s, demonstrated high specific and intra-specific variabilities, both in terms of the effects of cover crops on these pests and the responses of the latter to the various genotypes tested. Despite this, constants emerge for some species, although the results are also dependent on the race of nematode species, making assessment of the effect of catch crops on nematodes relatively complex. Some effects – those of Brassicaceae for example – seem to have been established: oil radish and white mustard are effective against beet cyst nematode, and brown and white mustards are effective against root-knot nematodes in leguminous crops. But not all Brassicaceae are equally beneficial against nematodes. For example, turnips have proved to be a major host for three species of *Meloidogyne* (root-knot nematodes). Other species, particularly among Poaceae, demonstrate interesting properties for the control of nematodes. But once again, in oats for example, differences in effects on nematodes exist between similar species or varieties. The main mechanisms involved in nematode reduction concern the non-host character of the plants (passive suppression) or their allelopathic character (active suppression).

The various species and trophic groups of nematodes in the soil are considered to be good indicators of its physicochemical properties and trophic networks. Yet, degradation of cover crop residues will follow different dominant microbial routes depending on their biochemical composition, and the associated effects of this on nematode communities appear to be greater than those associated with soil parameters such as temperature, moisture, pH and texture. Consequently, they favour the trophic groups that depend on them and this diversification is accompanied by a decrease in the dominance of phytoparasitic nematode species.

Micro- and mesoarthropods are favoured by cover crops and their residues, particularly when the residue mulch left on the surface is sufficient to modify the microclimate at the soil surface. The latter include mites that are predators of nematodes, leading to a reduction in phytoparasitic nematodes. Physicochemical characteristics and soil tillage types have a variable influence depending on the micro- and mesoarthropod group, with garden centipedes, for example, being very little affected by the soil tillage method but much more so by the introduction of various types of cover crop.

In terms of macrofauna, cover crops and their residues have very favourable effects on earthworm populations; this effect is even more significant if the field is managed using a no-tillage system. Earthworms accelerate the incorporation of residues into the soil and improve its physicochemical properties. It has also been shown that a high earthworm density is correlated with a decrease in pressure from certain pests, particularly phytoparasitic nematodes.

Slugs are some of the soil macrofauna pests most likely to be favoured by certain cover crops and their residues. However, major differences exist between cover crop types and sites.

Several species of Coleoptera, Diptera and Lepidoptera can have their larval stages – the crop pest stages – favoured by the use of a cover crop. However, it appears that marked biological activity of the soil reduces their populations, as a result of regulation by different antagonists. But these hypotheses are worthy of further exploration.

The dynamics of epigeic and aerial invertebrates are also influenced by the use of cover crops. In perennial crops, fallow period cover crops have demonstrated globally favourable effects on the diversity of generalist predator species and, as an indirect or direct consequence, unfavourable effects on the dynamics of numerous pests. In annual crops, the effects of the types – or even varieties – of cover crops on pests and natural enemies appear to be highly genotype-specific and screening of these against significant arthropods therefore helps to guide the choice of both cover crops and the main crops that are associated with or follow them, in order to minimize the risks of crop damage. This result is sometimes dependent on an increase in natural enemies, themselves potentially favoured by cover crops and their mulch (particularly weed seed predators). As with set-aside land planted with flowers, the potential benefit of which has been demonstrated for pollinators in decline in industrialized countries (but which are also criticized for using sometimes unsuitable non-native species), entomophilic angiosperm cover crops may represent an interesting option to promote flower-pollinating insects and restore effective pollination networks in agro-ecosystems, provided that they are plurispecific and attractive.

As far as vertebrates are concerned, cover crops considered to be favourable to wildlife tend to be those that are highly attractive to seed-eating birds. But, more generally, the majority of birds found in agricultural landscapes need to have access to areas of bare soil and will be disadvantaged by fields of cover crops with dense plant cover. As for micromammals, they are particularly encouraged by permanent plant cover in arboriculture, especially if it is based on leguminous crops, and research into repellent genotypes is ongoing.

Finally, a review of data concerning the relationships between cover crops and invertebrates or vertebrates highlights, first of all, shortcomings and significant needs in terms of research in order to make better use of cover crops to regulate pests (numerous pests and important natural enemies have not yet been the focus of any research in this respect) and, secondly, the fact that the known interactions reveal idiosyncratic behaviours (i.e. not fitting any rules – or at least any that have been elucidated – but, rather, with apparently individual characteristics), both generically and specifically. Improving our knowledge in this area may make it possible to avoid favouring pests by using certain cover crops and at the same time favour beneficial organisms for cash crops more effectively.

Most of the data available concerning the effects of crop volunteers on pests and natural enemies are based on observations or empirical information and the very few scientific studies that have been conducted indicate the absence of real risks for the same following crop.

2.4.4 The Difficulties Related to Integrating the Various Biotic Effects

For the three biological groups of organisms studied, the effect of using of cover crops still needs to be better understood, especially in terms of the various possible management options (which must be compatible with their primary function as catch crops), before they can be considered as an effective method for plant health improvement. Nevertheless, a few beneficial biological processes that are more general or already well-known (such as the increased biological activity of soils, promoting the diversity of nematode enemies and hence nematode regulation), as well as the allelopathic effects of several species and varieties of Brassicaceae or Poaceae on certain pests (biofumigation against soil-borne diseases and certain nematodes), could lead – more rapidly than others – to interesting applications in the management of cropping systems.

Alongside these positive aspects, which must be highlighted and, above all, examined in more depth to make their various applications more rigorous and more realistic on a large scale, it is also important to incorporate the negative aspects of cover crops in the design and management of cropping systems. The following points should therefore be highlighted: (i) cover crops almost systematically favour slug populations, (ii) some species of catch crops also promote phytopathogenic agents with partially saprophytic properties, and (iii) the effectiveness of cover crops against weeds, as well as pests and diseases, depends very significantly on agronomic parameters, as well as factors determining their growth and the type of soil tillage.

The analysis of the international scientific literature clearly demonstrates potential benefits, but also risks of failure, or even an exacerbation of plant health problems. These variable effects are primarily due to the specific and intra-specific characteristics of the organisms involved. The scientific analysis of the diversity of available information should therefore be carried out with caution and emphasize the high variability of effects, which must be taken into account. On the other hand, it appears that the French technical literature aimed directly at agricultural advisers and farmers primarily highlights the plant health risks of cover crops. These opinions, which are very probably mainly based on field observations, deserve closer examination and clarification in order to gain a more thorough understanding of the processes involved. However, it should be highlighted that in 2006 ("Cropping practices" survey) 48 % of sugar beet crops were already preceded by a catch crop such as white mustard, using "nematode-control" varieties. This illustrates that, in certain conditions, it is possible to combine a catch crop with a positive effect in terms of reducing certain pests and diseases.

In conclusion, this analysis simply reinforces the need to continue – and even step up – research in this area in order to ensure the best possible use of catch crops for plant health management, considering both their real risks and the real opportunities they offer.

2.5 Conclusion: A Rich But Incomplete Analysis of the Literature

The analysis of the literature helped provide an insight into the main biophysical mechanisms at work during fallow periods and to explain the value of cover crops and other techniques for the management of nitrogen on a cropping system scale. It also made it possible to describe the other functions of plant cover during fallow periods. However, the scientific literature does not cover a broad enough range of situations to be able to fully integrate either the variability of agricultural, soil and climate conditions in France or the numerous factors determining the effectiveness of catch crops and their effect on the crops that follow. The same is true for the other effects induced by cover crops, particularly with respect to crop pests and natural enemies. The scientific literature alone does not provide enough data to be able to rank or incorporate these effects.

Given this very marked variability, it is impossible to imagine an experimental setting that would be capable of producing references addressing all the questions raised in the various agricultural, soil and climate contexts. Consequently, the use of modelling is a relevant approach in order to examine the nitrate-trapping function of these crops and their effects on the following crop. About 1,300,000 different fallow period management situations have therefore been simulated using this approach. The results of this simulation work are presented in the next section.

References

Colbach N, Grancher S, Guyot SHM, Mézière D, Darmency H (2012) Changing agricultural practices modifies the species and trait composition of the weed flora. A simulation study using a model of cropping system effects on weed dynamics. In: Proceedings of the 6th international weed science congress, Hangzho, China, 17–22 June 2012

Dorsainvil F, Dürr C, Justes E, Carrera A (2005) Characterisation and modelling of white mustard (Sinapisalba L.) emergence under several sowing conditions. Eur J Agron 23:146–158

Chapter 3
Main Lessons Drawn from the Simulation Study

Eric Justes, Julie Constantin, Nicolas Beaudoin, Carolyne Dürr, Bruno Mary and Laurent Ruiz

3.1 Objectives of the Simulation Study Using a Dynamic Crop Model

The study of the literature relative to the impact of cover crops on nitrate leaching, and hence their primary role as "nitrate-trapping" catch crops in annual cropping systems, reveals a lack of data relating to certain regions in France. The same is true concerning their impact on other variables, such as the water balance or the yield of the next crop. However, since the planting of cover crops has become compulsory in all vulnerable zones since 2012, it is of the utmost importance to assess how effective they are in terms of reducing the nitrate concentration of drainage water, as a function of the soil and climate context in a broad range of French conditions. By

The original version of this chapter was revised.
An erratum to this chapter can be found at https://doi.org/10.1007/978-94-024-0986-4_5

E. Justes (✉)
UMR AGIR (AGroécologie, Innovations & teRritoires) 1248, University of Toulouse, INRA, INPT, Centre de Recherches INRA Occitanie-Toulouse, 24 Chemin de Borde Rouge, Auzeville, CS 52627, Castanet Tolosan 31326, France

UMR SYSTEM Univ Montpellier, Cirad, Inra, Montpellier SupAgro, CIHEAM-IAMM, 2, place Pierre Viala, Bat. 27, Montpellier 34060 Montpellier cedex 2, France
e-mail: eric.justes@inra.fr; eric.justes@cirad.fr

J. Constantin (✉)
UMR AGIR (AGroécologie, Innovations & teRritoires) 1248, University of Toulouse, INRA, INPT, Centre de Recherches INRA Occitanie-Toulouse, 24 Chemin de Borde Rouge, Auzeville, 31326 Castanet Tolosan cedex, Toulouse CS 52627, France
e-mail: julie.constantin@inra.fr

N. Beaudoin (✉)
UPR 1158 AgroImpact de Laon-Mons, Site de Laon, Pôle du Griffon, 180, rue Pierre-Gilles de Gennes, Barenton-Bugny 02000, France
e-mail: nicolas.beaudoin@inra.fr

E. Justes (ed.), *Cover Crops for Sustainable Farming*,
DOI 10.1007/978-94-024-0986-4_3

doing so, it will be possible to optimize the associated management technique and identify situations in which their use may have negative consequences on the next crop and, potentially, on groundwater recharge.

The simulation study conducted in addition to the analysis of the literature is carried out in a variety of different French soil and climate contexts and comprises three parts[1] (Fig. 3.1):

1. A central study concerning the impacts of fallow period management options on nitrogen (nitrate leaching, in particular), the water balance and the effects on the next crop.
2. A study concerning the sowing-emergence phase of cover crops, in order to identify problems likely to arise during this cover crop establishment phase, which is critical for its "nitrate-trapping" effectiveness.
3. A specific study concerning the autumn destruction phase of cover crops and the risks of soil compaction linked to mechanical destruction, as a function of the soil type considered.

This simulation study therefore aims to assess the impact of cover crops on the basis of the crop types used, management options (sowing and destruction dates) and their growing duration, primarily related to previous and subsequent crops, sown in the winter or spring (Fig. 3.2).

The situations selected enable simulation of a broad range of fallow period management options in regionally coherent cropping systems, climates and soils, thereby exploring a range representative of the diversity of existing situations. However the study does not claim to be exhaustive. The scenarios are selected, firstly, on the basis of climatic sites, soil types, crop rotations and residual mineral nitrogen content in the soil at the time of harvest of the previous crop and, secondly, of several types of cover crops, different ways of managing these cover crops and several previous crop residue management options.

[1] For the full results of these three simulation studies, see Chaps. 10, 8 and 9 of the report, respectively.

C. Dürr (✉)
INRA, UMR 1345 IRHS, 42 rue George Morel, Beaucouzé 49071, France
e-mail: carolyne.durr@inra.fr

B. Mary (✉)
INRA, UPR 1158 AgroImpact de Laon-Mons, Site de Laon,
Pôle du Griffon, 180, rue Pierre-Gilles de Gennes, Barenton-Bugny 02000, France
e-mail: bruno.mary@inra.fr

L. Ruiz (✉)
UMR SAS 1069, Université Bretagne Loire, INRA, Agrocampus Ouest-ESA,
65 Rue de Saint Brieuc, Rennes 35042, France
e-mail: laurent.ruiz@inra.fr

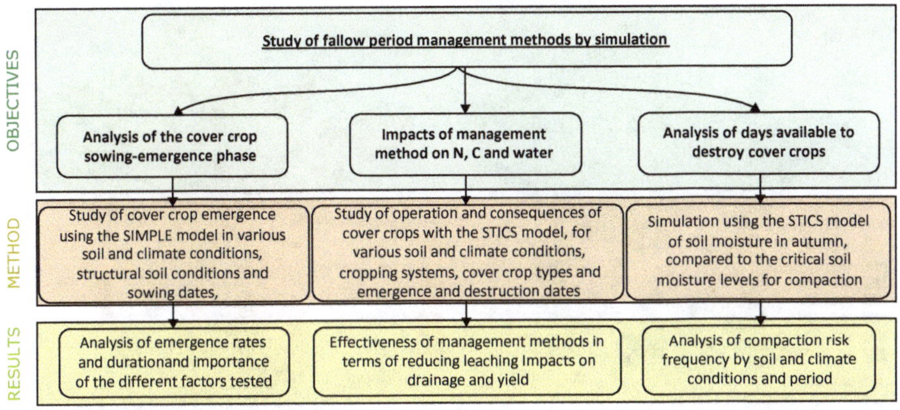

Fig. 3.1 Summary of the different parts of the simulation study, their objectives, general methods and the expected results

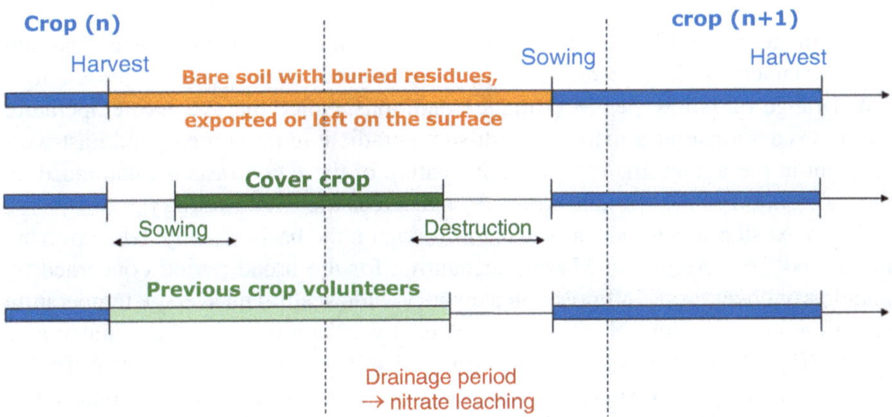

Fig. 3.2 Diagram of the different fallow period management options in an annual crop rotation with an impact on nitrogen, water and carbon balances

3.2 Methodology of the Three Parts of the Simulation Study

3.2.1 The Situations Simulated[2]

Climatic Sites

The first step was to perform climate zoning based on monthly data derived from the SAFRAN model (Météo-France) over the period 1993–2004 and on grids of

[2] For details of the situations selected for the simulations, see Chap. 7 of the report.

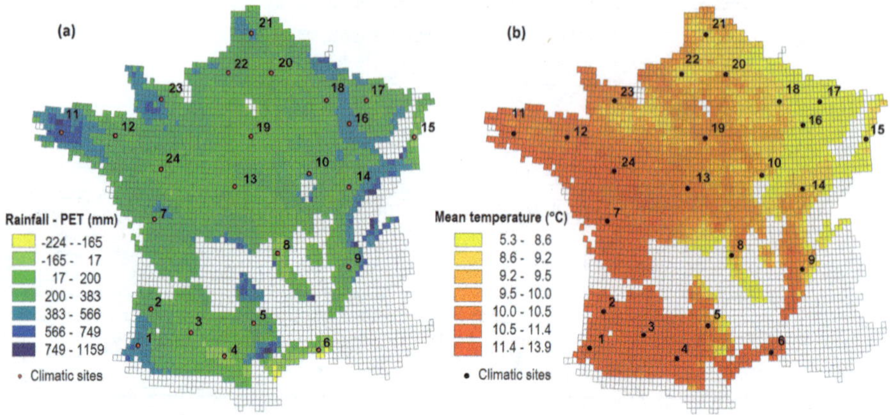

Fig. 3.3 Zoning by (**a**) (R-PET) balance and (**b**) average temperatures over the period August to March of the meteorological grids located in the main arable regions concerned by "Nitrate vulnerable zones" and location of the 24 climatic sites selected

approximately 12 × 12 km, covering the entire area of mainland France. The aim was to characterize its climatic variability and select contrasting sites representing a broad range of fallow period climatic conditions. Since the average temperature variations demonstrate a marked North-south gradient in the summer and East-west gradient in the winter, the representative nature of these contrasts is guaranteed by the distribution of the climatic sites selected across the whole country.

The next step was to perform climate zoning on the basis of daily data, covering the period from August to March, accounting for the broad period concerned by planting of cover crops in France on a country-wide scale. The average temperature variations retain a dual North-south and East-west gradient for the country as a whole (Fig. 3.3a). The water deficit (rainfall minus potential evapotranspiration (PET)) over the period August to March demonstrates much greater regional variation (Fig. 3.3b). This characteristic was taken into account when selecting sites.

Ultimately, for each of the 24 sites selected, 20 climatic years, from 1988 to 2008 (SAFRAN model data), are considered. It should be noted that the sites were selected to be representative of all the contrasts found in French arable system regions (excluding mountain regions). However, the objective was not to obtain a spatially weighted representation of these contrasts as a function of arable land surface area. The primary objective is to analyse the effect of catch crops for the whole range of soil and climate conditions found in France.

Soils

For each climatic site and its surrounding agricultural region, 2–4 representative sites in terms of available water capacity (AWC), depth and surface texture were selected using the French soil database on a scale of 1/1,000,000. The selection of soils did not aim to be exhaustive but to provide a relatively broad range of soil

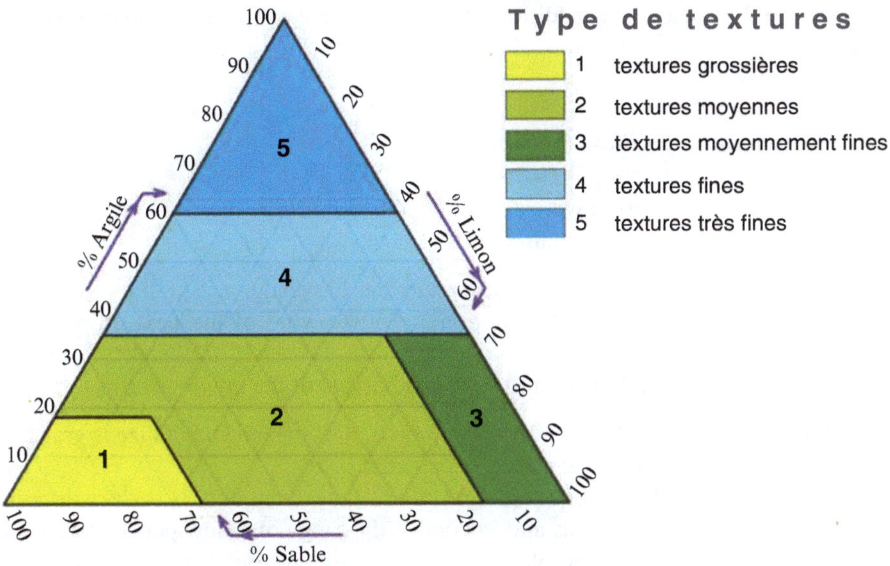

Fig. 3.4 Soil texture triangle (CEC, 1985)

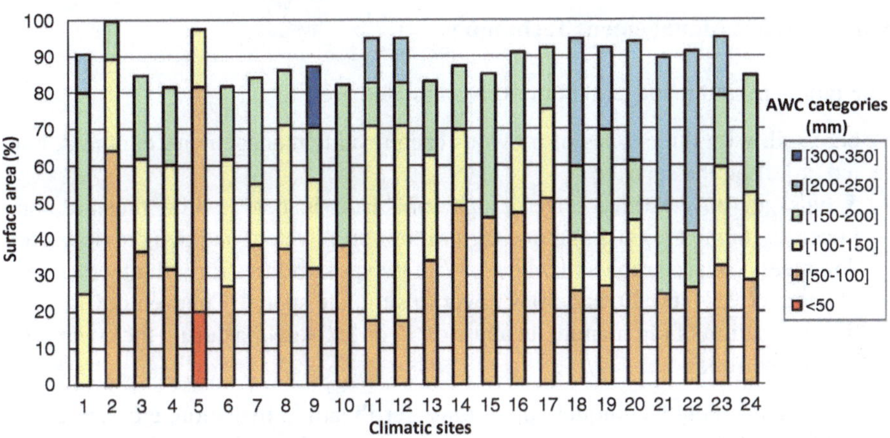

Fig. 3.5 Cumulative surface area of French soil types represented by their AWC category (expressed in mm) associated with each climatic site in Fig. 3.3

types characterized by contrasting depths and AWCs, representative of the zone considered (Fig. 3.4).

The soils selected belong to 4 of the 5 classes in the texture triangle (Fig. 3.5): rough (sandy soils), medium (loams), medium-fine (clay-loam soils) and fine (clay soils).

The very fine texture (clay content > 45 %) is not represented since it corresponds to very specific heavy clay soils, with physical and hydrodynamic properties that are difficult to take into account, particularly in crop simulation models such as STICS (see Sect. 3.2.2, below); in addition, these soils account for few arable cropping systems.

Crop Rotations and Fallow Period Types

Three types of fallow period were defined on the basis of the the previous-next pairing (previous main crop-next main crop):

- Wheat-maize: long fallow period after summer harvest;
- Oilseed rape-Wheat: short fallow period with sowing of the next crop in the autumn;
- Maize-Maize: long fallow period with late harvest of the previous crop.

The wheat is winter wheat – either common or durum – and the maize is grain or silage maize, depending on the region considered.

These three typical rotations represent, first of all, a significant surface area in French arable cropping systems and, secondly, the range of fallow period durations for which a cover crop can be planted. Finally, they are found in every region of France, although their surface areas vary.

Fallow Period Management Techniques

The fallow period management options simulated are:

- bare soil, with three types of previous crop residue management: exported, buried or left on the surface;
- 5 types of cover: three cover crops (one Brassicaceae: white mustard, one Poaceae: Italian ryegrass and one leguminous crop: vetch) and two previous crop volunteers (oilseed rape and wheat). With two possibilities for volunteers: dense uniform cover (100 %), and 50 % cover to simulate spatially uneven cover;
- the application of pig manure, at a rate of 75 or 150 kgN/ha during the cover crop sowing period.

The management techniques (ploughing, fertilization, irrigation, etc.) are determined for each type of crop on the basis of the regional practices analysed as being dominant in the 2006 "Cropping practices" survey. Three hypotheses for residual soil nitrogen content at harvest were selected: initial statuses N1 (20 kgN/ha), N2 (60 kgN/ha) and N3 (100 kgN/ha), representing the variability of situations that might result from the more or less careful determination by farmers of the optimal dose of nitrogen fertilizer for the previous crop.

For cover crops, six emergence dates (from July to October) and six destruction dates (from October to March) are tested in the simulations.

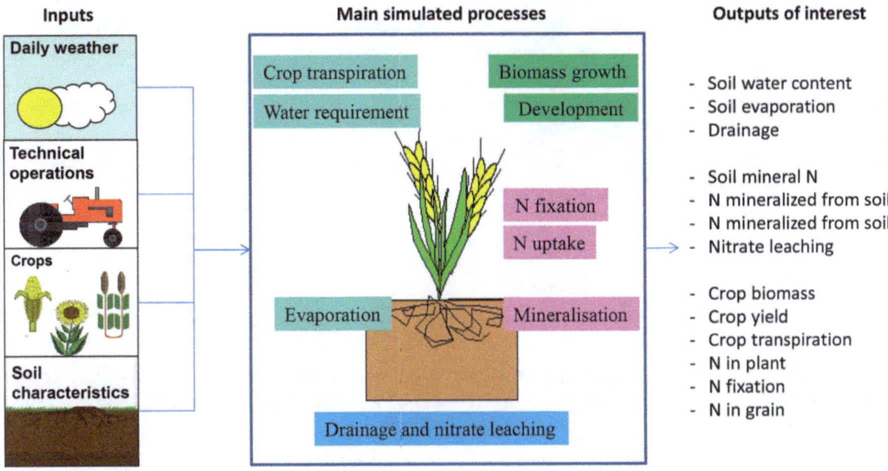

Fig. 3.6 Diagram showing the input data of the STICS model, the principal processes simulated and the output variables of interest for the study

3.2.2 *The Crop Model Chosen: STICS (Version 6.9)*

Simplified Description of the STICS Model

The STICS[3] soil-vegetation-atmosphere model is one-dimensional and incorporates the variabilities and diversity of the climate, soil and technical practices. It simulates water, carbon and nitrogen balances on a daily time-scale (Fig. 3.6) and manages the individual or combined effects of water and nitrogen stresses. Abiotic stress factors other than water and nitrogen are not taken into account by the model, and neither are biotic stress factors (diseases, pests).

STICS simulates yield variables (quantity and quality) and environmental variables such as nitrate leaching. Its primary value is that it is capable of simulating the consequences of environmental and cropping system variations on production, as well as the environmental impact of this production on water, nitrogen and carbon.

The crop is assessed overall by its aerial biomass and its nitrogen content, its leaf area index and number, the biomass and nitrogen content of the plant parts harvested. The soil is described as a succession of horizontal layers, each characterized by its water content and its mineral and organic nitrogen contents. Interactions between the soil and the crop are ensured by the roots, defined by the root density distribution in the soil profile. Growth of the crop is driven by the carbon balance, in which the solar radiation intercepted by the foliage is converted into aerial biomass by photosynthesis. The nitrogen balance of the crop is simulated in parallel with the carbon balance, upon which it is partially dependent, with the biomass being simulated as a function of the nitrogen status of the plant and the availability of mineral nitrogen in the soil.

[3] Brisson et al. (1998)., Brisson et al. (2008).

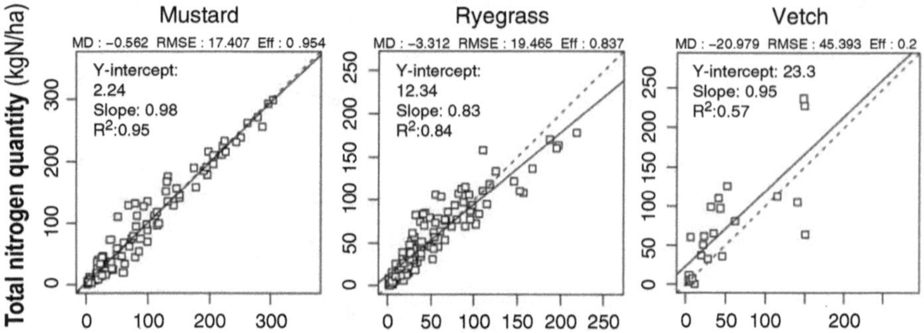

Fig. 3.7 Comparison of nitrogen uptake by the 3 types of cover crop (simulated values on the x-axis and measured values on the y-axis). MD = Mean deviation, RMSE = Standard error, Eff = Efficiency of the model

Predictive Quality and Validity Range of the STICS Model

Prior to conducting this simulation study, the predictive quality of the STICS model was assessed and its validity range determined in order to simulate both arable cropping systems and cover crops, and ensure that the predominant processes in the carbon, water and nitrogen cycles were correctly simulated. As an illustration, the comparison made during the study's initial validation phase between experimental data and results of simulations concerning the nitrogen uptake by the cover crop is presented (Fig. 3.7). The assessments of the predictive quality also concerned the water balances and mineralization of catch crop residues. All these "validations" indicated that the simulations were precise enough to be considered reliable.

The STICS model is parametrized for the main cultivated crops in temperate countries,[4] including wheat, maize and oilseed rape, as well as for mustard and Italian ryegrass cover crops.[5] The overall level of precision of the simulation is an error of around 15–20 % for nitrogen uptake, biomass and yield variables. The precision is generally greater for the water balance and water content of the soil (error is often less than 10 %), but it is poorer for the soil's mineral nitrogen content (generally between 15 and 25 %) since numerous processes (physical, chemical and biological) all occur simultaneously.

Furthermore, the STICS model was re-parametrized for vetch in the context of this study, using unpublished data supplied by the the INRA's Toulouse unit (UMR AGIR) and supplemented by data from Arvalis-Institut du Végétal and CETIOM. Additional parametrization was performed to correctly simulate oilseed rape and wheat volunteers on the basis of the original parameter settings for these two main crops.

Finally, a comparison of the results obtained from the simulated situations together with experimental data used to parametrize and evaluate the STICS model ensured that the model was not used to perform simulations outside its validity range.

[4] Brisson et al. (2008).

[5] Dorsainvil F. (2002).

The results were satisfactory and the model could therefore be used with confidence to perform "numerical modelling". However, since modelling of certain conditions was inadequately validated, these were not the subject of simulations: this was the case with undersown catch crops, mixed catch crops (legumes and non-legumes), and soils with a high clay content (>45 % clay).

3.2.3 *Evaluation of Days Potentially Available in the Autumn for Destruction of the Cover Crop*

The number of days potentially available for the mechanical destruction of the cover crop in the autumn was evaluated on the basis of characterization of the physical condition of the soil. The bearing capacity of the soil, linked to its texture, density and moisture content, was calculated daily over this autumn period in order to determine the days available for destruction of cover crops in good conditions. This evaluation was performed on the 24 climatic sites and for the different soil types. Hence the study incorporated a broad range of soils, ranging from clay soils (37–43 % clay) to sandy ones (>80 % sand). The simulations were performed for the 20 climatic years.

For each soil type, a critical moisture level per soil horizon (0–10 and 10–30 cm) was calculated as a function of the texture and water content at field capacity (or water-holding capacity) of the soil, as well as the stresses exerted by agricultural machinery during the mechanical destruction of the cover crop (40 kPa). For agricultural machinery stresses of 40 kPa, these moisture levels range from 100 to 139 % of the water content at field capacity, with critical levels that decrease as the clay content of the soil increases.

The daily simulation of the water content of the upper two soil horizons, calculated using the STICS model, is then used as a criterion to evaluate the bearing capacity of the soil and deduce the compaction risk day by day. A day is therefore considered to be available if the moisture contents of each of the two horizons are below the critical level for the horizon. If the condition is not met for at least one of the soil layers, the day is considered to be "at risk" for compaction and hence "unavailable".

3.2.4 *Catch Crop Sowing-Emergence Phase Simulated Using the SIMPLE Model*

Characterization of the SIMPLE Model

SIMPLE[6] is a model that simulates the germination and emergence of different crops. The model therefore makes it possible to predict the rate and speed of germination and emergence. Its input variables (Fig. 3.8) characterize the crop types

[6] Dürr et al. (2001). This article has been cited 33 times (source: Web of science).

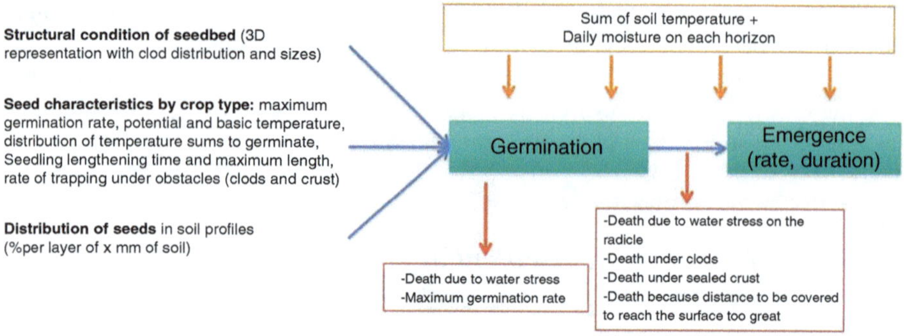

Fig. 3.8 Simplified diagram showing how the SIMPLE model works

studied and define the environmental conditions in the seedbed (temperature and water content) as well as the effect of soil tillage and passage of the seed-drill in a given soil (size and spatial organization of aggregates, position of seeds).

This model is stochastic, i,e. each simulation is performed by modelling the fate of 1000 seeds sown. The seeds are randomly distributed in the soil layers using the probability distribution of sowing depths in the soil according to the type of sowing techniques. This distribution depends on the type of soil tillage and the seed-drill used. Several reasons for non-emergence are simulated: non-germination, mortality of the seedling due to water stress on the root following germination, trapping under a clod or sealed crust, insufficient seedling length to reach the surface. The effects of pests and diseases are not simulated.

Of the 24 sites selected for STICS simulations, 9 were selected for this sowing-emergence phase study, to represent the range of different classes of (rainfall-PET) balance and average temperature during fallow periods. The sites selected were sites 2, 4, 6, 10, 11, 15, 20, 21 and 24 (see Fig. 3.3).

The simulations were performed for the three types of cover crop (mustard, Italian ryegrass and vetch) and three sowing dates (15/07, 15/08, 15/09). The simulations were taken to the following 10 November. Three structural seedbed conditions were selected: fine, medium and coarse. These conditions differ in terms of the proportion of clods of various dimensions and their spatial organization in the seedbed (deep/surface). Three types of planting were simulated: broadcast sowing without soil tillage, with the seeds being covered with chopped straw; broadcast sowing after surface stubble ploughing, sowing using a cereal seed-drill after ploughing and harrowing to prepare the seedbed. The simulations take into account the interaction between the structural condition of the soil and seed distribution profile in the seedbed.

Validity Range of the SIMPLE Model

The SIMPLE model was validated on field trials on cover crops for various soil and climate types and sowing methods.[7] It is nonetheless important to specify the validity range of the model. First of all, the seeds are considered to be in sufficient contact with the soil to be able to germinate when the soil moisture content is greater than the water potential for seed germination; situations of poor soil-seed contact ("hollow" soils) are not simulated. Likewise, the impact of very high temperatures (>25 °C average temperature, or maximum temperatures of over 38 °C which prevent germination) are not taken into account in these simulations performed using the SIMPLE model. It is also necessary to point out that the effects of biotic stress (diseases and pests) on emergence rates are not simulated by this model. In general, little data is available concerning these effects, either in the scientific literature or the technical documents. Consequently, the impact of biotic limiting factors remains difficult to estimate.

3.2.5 Analysis of the Simulations and Presentation of Results

Analysis of the Simulations

The simulations are analysed with a view to evaluating the sensitivity of an output variable (simulated) to input parameter variations (climate, type of cover crop, management option, etc.). The objective is to identify the dominant factors of the effect of different fallow period management options (including cover crops) on water and nitrogen balances from harvest of the previous crop to harvest of the next crop. The simulations do not take into account the cumulative effects of "routine" catch crop insertion in the long term; these effects are only covered by the analysis of the literature (see Sects. 2.2.3 and 2.3).

The numerous simulation repetitions, in particular over the 20 climatic years, make it possible to consider the frequency of occurrence of a result (a seed emergence rate > 75 %, a leaching rate < 25 kgN/ha, for example) as a probability of obtaining this result and hence to quantify the risks of low catch crop effectiveness as a function of conditions or practices.

The simulated situations are selected to be representative of the diversity of conditions found across the country and hence the range of French conditions, but not their frequency of occurrence. As a result, the mean of the values obtained on the 24 sites (9 sites sites for the sowing-emergence phase simulations) is not a "France" mean, the calculation of which would assume knowledge of the "weight" of each situation (climatic, soil, etc.); consequently, the results are not expressed as an aver-

[7] Dorsainvil F et al. (2005).

age over the 24 sites, but the between-site variability is analysed in order to assess the response variability of the effect of cover crops. Neither do the simulations make it possible to perform zoning on the basis of the 24 sites alone, which would involve plotting "frontiers" and linking a simulated climatic site to a given region.

Representation of the Variability

Given the extent of the variability of the phenomena, related particularly to the variability in climate, the results of the simulations are represented as a boxplot (also known as a box-and-whisker diagram or plot). This makes it possible to represent most of the population of the results obtained and is therefore more informative than the mean associated with its variance.

Boxplots display the distribution of the variable considered according to the following rules (Fig. 3.9). The median (line in bold) represents the "middle" of the distribution. This means that 50 % of the observations made have a value above this median and 50 % have a value below it. The rectangular box corresponds to 50 % of the distribution centred around the median (25 % above the median, 25 % below). The dotted lines ("whiskers") extending from the top and bottom of the box end with dashes: 95.3 % of the distribution is found inside these. The circles (here) or dots located beyond this correspond to the extreme values, which therefore account for less than 5 % of cases.

This graphical representation therefore makes it possible to depict the distribution and thus the response variability of the variable, for a set of given conditions. It offers the advantage of showing the central trend, as well as the variability and the asymmetries of extreme values, as a function of the conditions studied.

Fig. 3.9 Distribution of values in a boxplot

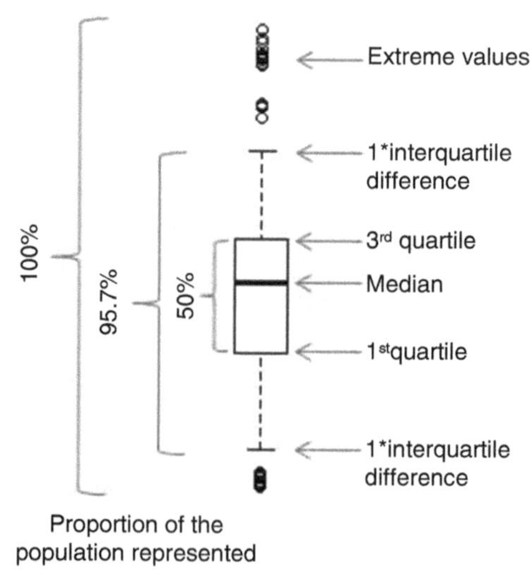

Quality Control of Simulations and Precautions for Use

The predictive quality and validity range of the models used were specified (see Sects. 3.2.2 and 3.2.4) before performance of the actual simulations. However, the simulations do not fully replace experimental data, but, rather, supplement them, specifying magnitudes and directional variation trends for the effect of catch crops, with a precision of around 5–10 % for the water balance and 15–20 % for the nitrogen balance. Although, for convenience of language, the results of the simulations are presented as statements, it must be borne in mind that these results correspond to estimations of reality. However they are precise and consistent enough to be presented in this report.

These numerical results represent new data. Ideally, it would have been preferable to be able to compare these simulation results with real situations. This would require extremely complex but valuable additional studies.

3.3 Generic and Specific Lessons Drawn from the Simulation Study[8]

This chapter contains the conclusions or synoptic data that, firstly, apply to a large number of soil and climate conditions and are therefore relatively generic in nature and, secondly, concern more specific situations for the inclusion of catch crops in cropping systems or particular management methods. Twelve main lessons can be drawn from this simulation study.

1. **Catch crops can be successfully planted in a broad range of situations, as long as the sowing date is adapted to the soil and climate conditions**

The catch crop planting simulation study was conducted sequentially, using the STICS model to predict the water content of the soil in the seedbed on the basis of the "current" climate, and then the SIMPLE model to predict emergence (date and rate). Generally, the results of the simulations indicate that the emergence of catch crops is potentially sufficient for planting in current summer conditions to be acceptable (emergence rate > 75 %; Fig. 3.10), both in the case of very simplified planting techniques (under a straw mulch) and in a seedbed following soil tillage.

Cases of poor emergence (emergence rate < 50 % of seeds) occur in various climate conditions, but with a relatively low average frequency: less than 1 year in ten for Northern French sites, and 2 years in ten in the South-Western climate. In the latter case, sowing on 15 August and 15 September reduces the risks of non-emergence compared to early sowing on 15 July. In Mediterranean climates, emergence demonstrates a high level of year-to-year variability, with low emergence rates (<25 %) in around 2 years in ten, and almost total failure (emergence

[8] This section and subsequent sections are based almost entirely on Chap. 11 of the report dedicated to the study conclusions.

Fig. 3.10 Emergence rate frequencies for the three sowing dates at the 9 sites over 20 years (*red* circle)
Meaning of emergence rate classes:
≥75 %: estimated rate sufficient to obtain effective cover with the cover crop;
between 75 and 50 %: the plant density may be sufficient but an uneven spatial distribution prevents optimal growth and mineral nitrogen absorption;
between 25 and 50 %: the plant density is insufficient, particularly since the distribution of emerged plants is uneven;
<25 %: rate corresponding to total failure of sowing

rate < 10 %) in 2 % of simulations, irrespective of the sowing technique (with or without soil tillage). These failures occur particularly with early sowing (sowing on 15 July); consequently, this date is, overall, unsuitable in the "dry" climates of Southern France, where late sowing is more effective (sowing on 15 September).

Practically speaking, over and above the need to adapt the planting technique (which involves agricultural machinery, in particular) on the basis of the previous crop and the physical and hydric properties of the soil, 10-day weather forecasts (which are increasingly reliable) could be used to adapt the sowing date and method (type of machinery and sowing depth) to the moisture conditions of the soil's surface horizon, thereby maximizing the chances of success of catch crop planting.

It is important to remember that the methods used for the simulation study do not take into account potential losses due to biotic effects such as pests and diseases (fungal diseases, bird damage or slug and snail damage). Yet, significant losses may occasionally be incurred in the event of major pest or pathogen attacks (particularly since catch crop seeds are not generally treated). However, emergence losses due to pests and disease still need to be assessed in real conditions for a broad range of field situations given the sparse nature of currently existing data.

2. **Catch crops reduce drainage but have no effect on the water supply for the next crop if their destruction date is well chosen**

Drainage (transfer of water beyond the soil) is what drives groundwater recharge and leaching (carrying of solutes such as nitrate by vertical transfer) and hence nitrate losses beneath the rooting layer of main crops.

Catch crops almost systematically reduce annual drainage (amount of water carried below the soil). The median reduction ranges from 20 to 50 mm, depending on

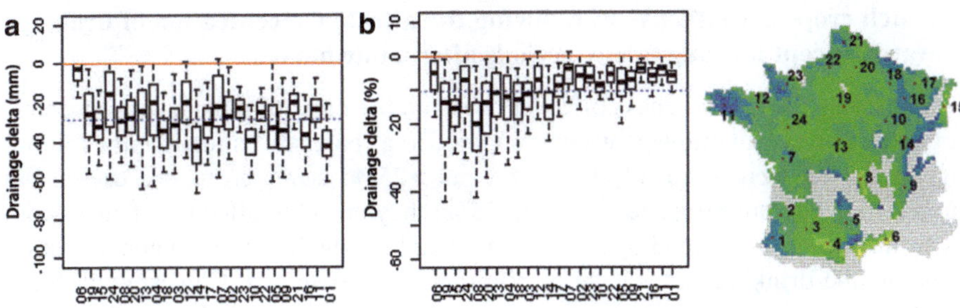

Fig. 3.11 Effects of cover crops on drainage Drainage differential (or delta) between mustard and bare soil, in a long fallow period (wheat-maize), as a function of climatic site (*blue line* = inter-site average) in millimetres (**a**) and percent (**b**). [Mustard emerged on 25/7, 10/8 or 15/8, and destroyed on 10/11]

the soil and climate conditions as well as the cover growth level. It corresponds to an additional water consumption compared to bare soil of 5–15 mm per tonne of catch crop biomass, which is variable depending on daily potential evapotranspiration and rainfall distribution. This cover crop transpiration is greater than the reduction in soil evaporation due to the plants. Ultimately, catch crops induce an actual median additional evapotranspiration of 20–50 mm, which corresponds to the drainage reduction level.

For destruction dates from early October to early March, this reduction in drainage is not generally accompanied by a reduction in soil water content at the end of the autumn or start of the spring (mid-April). This is because winter rainfall levels easily compensate for the evapotranspiration of the cover crop and enable restoration of soil moisture (in all the soil and climate conditions simulated) between destruction of the catch crop and sowing of the next main crop. Since available water capacities appear to be little affected or not affected at all, generally there is no negative impact on the next crop as a result of this additional water consumption by the catch crop. However, if the catch crop destruction date is too late (late March to mid-April) or in the event of very low winter rainfall, the soil water reservoir might not be replenished because renewed growth of the catch crop in the spring might then lead to significant evapotranspiration. This might not be compensated for by the rainfall and hence prevent the replenishment of the soil reservoir for the next crop (Fig. 3.11).

This reduction in drainage is less than 10 % of the water drained in the great majority of soil and climate sites, but can be as much as 25 % in some years at sites with low winter rainfall. Consequently, in regions where there is extensive use of catch crops, these may potentially reduce groundwater recharge in hydro-geological situations where this depends mostly on under crop-based drainage. In this case, an assessment of the impact that the extensive local use of catch crops might have on groundwater recharge is necessary in order to analyse the benefit of the expected reduction in nitrogen concentration compared to the cost or risk with respect to recharge.

3. **Catch crops are effective in reducing the nitrate concentration of drainage water, except for long fallow periods after a late harvest**

To qualify the effectiveness of cover crops on the basis of the nitrate concentration reduction rate in drainage water compared to a "bare soil" situation, the following thresholds were used: (1) reduction rate > 25 %: cover crops are useful, (2) rate > 50 %: they are effective, (3) rate > 75 %: they are very effective. This assessment may be supplemented by a criterion related to whether or not a concentration of below the drinking water level of 50 $mgNO_3/l$ is obtained. However, since the nitrate content of drainage water depends on the initial amount of mineral nitrogen in the soil (at harvest of the previous crop), catch crops may be effective – or even very effective – without always guaranteeing that the drainage water meets drinking water standards; in other words, catch crops cannot mitigate the polluting effect of excessive mineral nitrogen in the soil and do not therefore exonerate farmers from implementing the strict and appropriate management of nitrogen fertilizer application to crops.

Overall, if emergence is sufficient and the cover dense and uniform enough, the effectiveness of non-legume catch crops is indisputable for reducing leaching (amount of nitrate lost) and drainage water nitrate levels (Fig. 3.12) in both short fallow periods (e.g. an oilseed rape-wheat rotation) and long fallow periods (e.g. a wheat-maize rotation). However, this conclusion is not valid for long fallow periods after a late harvest (e.g. a maize-maize rotation, see point 5 below).

Hence, the study indicates the effectiveness of non-leguminous catch crops across all of the French climatic sites tested: the nitrate concentration reduction rate in drained water is generally more than 50 % (and may exceed 75 % for a large number of situations with high precipitation) on the optimum emergence and destruction dates, which can differ depending on the soil and climate. This conclusion is valid even if the growth or water and nitrogen nutrition levels of the catch crop are not optimal, since the levels of biomass produced and nitrogen uptake depend on the availability of water and mineral nitrogen during the fallow period.

While, on a scale of several years, catch crops are always useful and are effective for reducing nitrate leaching in the very great majority of situations – even at soil and climatic sites in which the median value of drainage is low – it is clear that if no drainage occurs in a given year (including under "bare soil"), the catch crop will not have been useful. The impact of catch crops must therefore be analysed for a period lasting several years in order to reliably assess their effectiveness. This assessment was made taking into account the climate of each site over the last 20 years.

More specifically, it is possible to draw the following conclusions (Fig. 3.12):

- The level of effectiveness in terms of reducing the amount of nitrate leached and, consequently, the nitrate concentration of drainage water, depends on climatic factors: the higher the rainfall is at a site, the more effective catch crops are at trapping mineral nitrogen and hence reducing the amount of nitrate transferred underground, while the reduction in drainage volume is low.
- The effectiveness of catch crops is increased in long fallow period situations since the duration of the cover crop growth period can be greater, maximizing nitrogen trapping. In long fallow periods, the median nitrate concentration reduc-

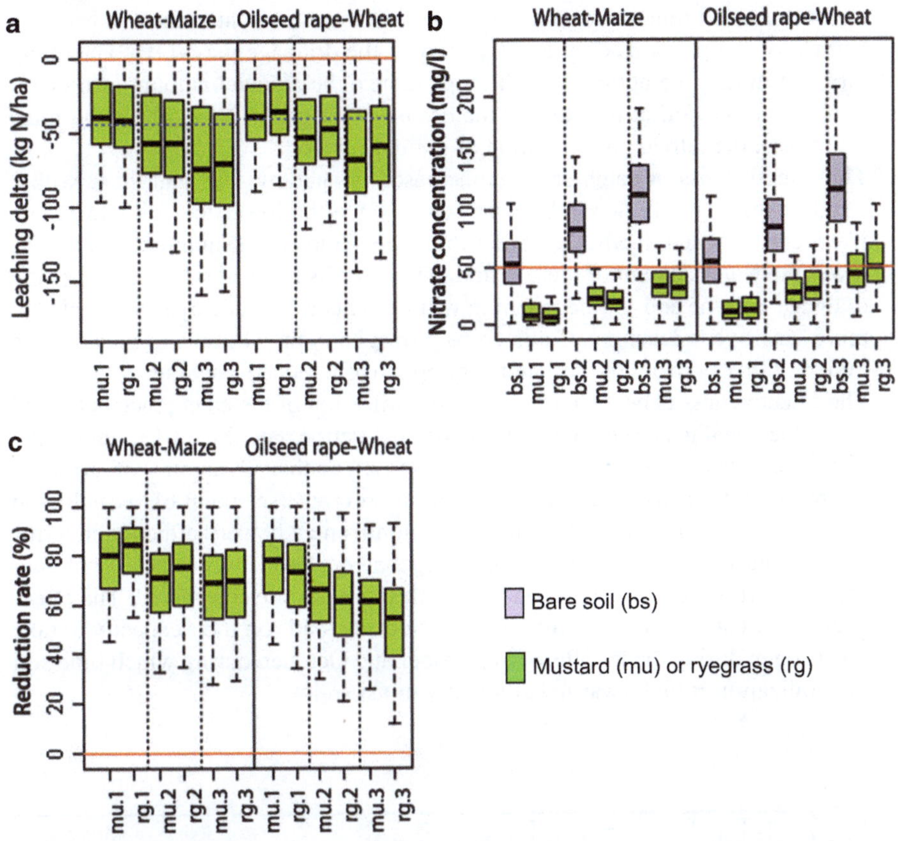

Fig. 3.12 Effects of different types of cover crop on nitrate leaching for two types of crop rotation
Difference (or delta) in nitrate leaching (**a**), nitric concentration (**b**) and concentration reduction level (**c**) by type of cover crop, depending on initial soil mineral nitrogen status (1 = 20, 2 = 60, 3 = 100 kgN/ha) and crop rotation (*blue line* = average per rotation)
[Simulations on all sites; for emergence dates at 25/7 and 10/8, and destruction dates at 1/10 and 20/10 for the oilseed rape-wheat rotation and at 10/11 and 10/2 for the wheat-maize rotation]

tion rate in drainage water is lower (around 50 %) at low-drainage climatic sites (Mediterranean zones or sites in central France), which represent the least favourable sites in terms of catch crop effectiveness. This median level may rise to as much as 90 % for the sites with the highest rainfall (oceanic sites in the South-West and North-West of France).

- The effectiveness of catch crops – simulated in both short and long fallow period situations – in terms of reducing the nitrate concentration of drainage water has been established for the three initial mineral nitrogen levels, i.e. even when the initial residual mineral nitrogen content in the soil is low (initial status N1 in the simulation corresponding to an "irreducible" residual amount of 20 kgN/ha over a depth of 90 cm). This result can be explained by the fact that soils mineralize humified organic matter and "naturally" produce mineral nitrogen during the summer and autumn, increasing the mineral nitrogen content of the soil between harvest of the previous crop and the start of the drainage period. By absorbing nitrogen during the autumn, catch crops reduce the residual amount of mineral nitrogen in the soil at the start of the drainage period (in November) and thus, ultimately, the nitrate concentration of drainage water.

- The effectiveness of catch crops is increased in shallow soils and those with a rough texture (sandy or sandy loam soils), since the lower the soil's available water capacity and depth, the earlier drainage begins in the autumn. Hence, the lower the water content of the soil and the higher its renewal rate, the higher the drainage volume and nitrate leaching will be. This explains why sandy soils are more sensitive to leaching. Catch crops grown early may be able to quickly trap mineral nitrogen from the soil and very effectively reduce leaching.
- The effectiveness depends more on the optimization of the emergence date (and hence the sowing date) than it does on the destruction date (Fig. 3.13), unless the post-emergence growing time is very short (less than 2 months of growth). Therefore, putting aside the case of leguminous crops (see point 6), a catch crop that has had a growth period of at least 2 months may be very effective in terms of reducing nitrate leaching, where emergence occurs between the end of July and the start of September, depending on the climatic site considered. The effectiveness of catch crops in short fallow periods depends on their capacity to take up nitrogen during the months of September and October, during which time soil mineralization is high (warm, sufficiently moist soils).

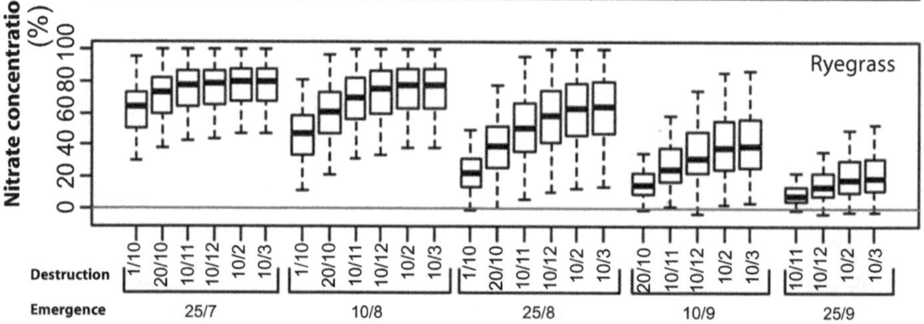

Fig. 3.13 Effect of emergence and destruction dates. Nitrate concentration reduction rate compared to reference bare soil as a function of emergence and destruction dates of mustard and ryegrass. [Simulations on all sites, for an initial mineral nitrogen content of 60 kgN/ha in the wheat-maize rotation]

4. **Legumes can be useful for reducing nitrate leaching, although they are only about half as effective as non-leguminous species (Brassicaceae and Poaceae)**

Overall, the simulations show that mustard (Brassicaceae) and Italian ryegrass (Poaceae) have comparable effectiveness in terms of reducing leaching and the nitrate concentration of drainage water for exactly the same emergence dates (Fig. 3.14). However, in certain conditions, mustard is more effective than ryegrass for trapping nitrogen. This greater effectiveness is especially apparent when the growing time is short (short fallow period or late sowing in September) or when the soil is deep (>90 cm and with a high available water capacity); in these situations, due to their faster development and better deep rooting capacity, Brassicaceae are more effective at reducing nitrate leaching.

The simulations indicate that vetch presents a median effectiveness that is about half that of mustard and ryegrass in terms of reducing leaching and nitrate concentrations in drainage water (Fig. 3.14). Since it primarily obtains its nitrogen through symbiotic nitrogen fixation, this legume therefore takes up only part of the available mineral nitrogen in the soil, in contrast with non-leguminous species such as mustard or ryegrass. This result confirms the fact that leguminous crops are not the most effective in terms of reducing nitrate concentrations. It does, however, demonstrate their usefulness in achieving nitrate concentration reductions comparable to those achieved with uneven wheat volunteers, although they have a similar impact on drainage to that of non-leguminous catch crops. Consequently, in terms of nitrate leaching it is better to plant a vetch cover crop (or legume) during the fallow period than to leave the soil bare with no cover crop.

One solution could be to sow a mixture of crops including a legume, or only a legume that would grow mixed with volunteers, to avoid the phenomenon of pre-emptive competition, particularly for nitrogen.

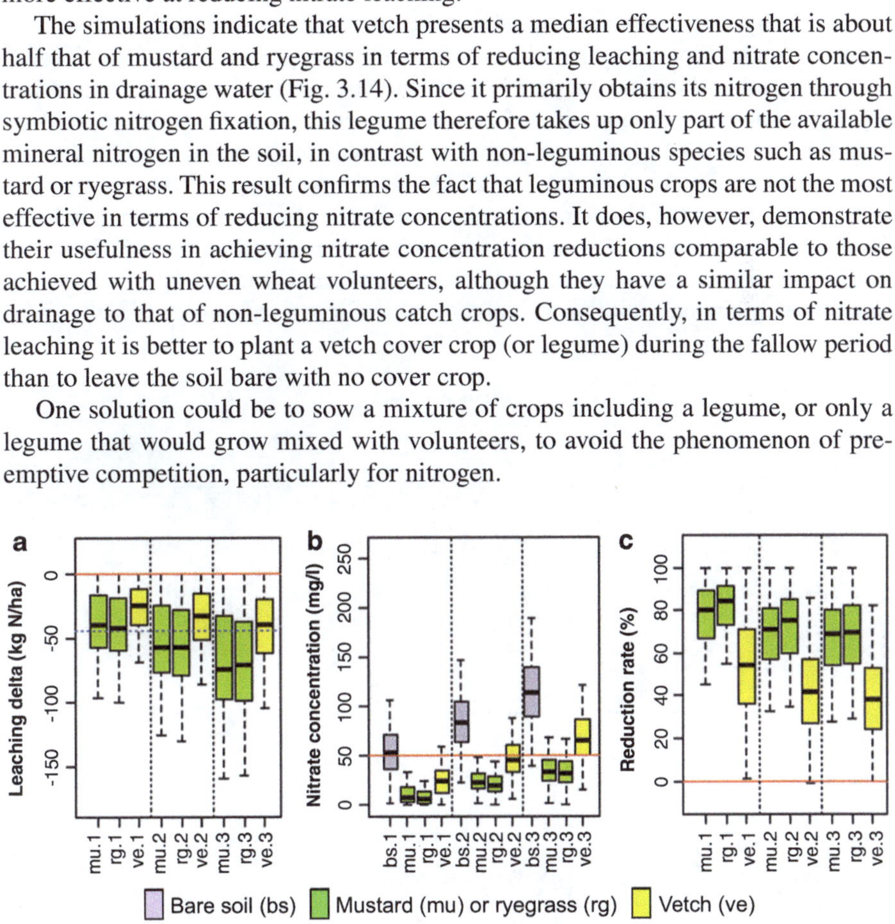

Fig. 3.14 Effects of a legume cover crop on reducing nitrate leaching compared to other families of catch crops. Differential (or delta) for nitrate leaching (**a**), nitric concentration (**b**) and concentration reduction level (**c**) by type of cover crop, depending on initial soil mineral nitrogen status (1 = 20, 2 = 60, 3 = 100 kgN/ha) and crop rotation (*blue line* = average). [Simulations on all sites; for emergence dates at 25/7 and 10/8, and destruction dates at 10/11 and 10/2]

5. **Catch crops are ineffective after grain maize and their effectiveness is relatively low after silage maize: this requires very strict management of nitrogen fertilizer application to avoid nitrate leaching in such cases**

After silage maize, catch crops are always useful for reducing leaching and hence nitrate concentrations in drainage water (Fig. 3.15), with a median reduction of 35 % for mustard and only 25 % for Italian ryegrass. In simulated situations representative of the Southern half of France, this reduction in concentration is around 50 % (threshold used to qualify the effectiveness of the catch crop) since catch crop growth may be significant for sowing dates from mid to late September (emergence simulated at 15 September), and the exporting of maize stover enables sowing in optimal conditions and total access to light for the plants.

Conversely, the study indicates that following grain maize harvested in mid-October, catch crops are generally insufficiently effective (and this effectiveness decreases further if the harvest date is later) and not very useful for reducing leaching and the nitrate concentration of drainage water (median nitrate concentration reduction rate < 10 %), since their growth is too weak in the autumn due to the late sowing date (emergence simulated at 25 October in the study). However, it can be noted in the case of the Mediterranean site (site 6) that mustard significantly reduces the nitrate concentration of drainage water, with a median reduction of 30 %.

This result reveals that, in the case of a maize-maize rotation (or spring crop), particularly grain maize, it is essential that the application of nitrogen fertilizers be

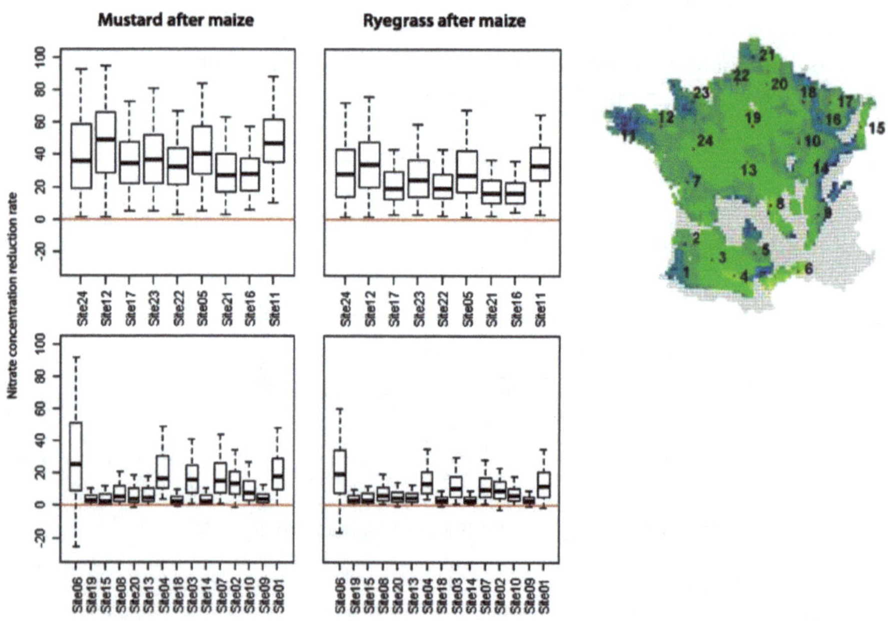

Fig. 3.15 Nitrate concentration reduction rates with cover crops (mustard and ryegrass) as a function of sites in maize-maize rotations, silage (*top*) and grain (*bottom*)

very strictly managed to ensure that the residual mineral nitrogen content of the soil when the maize is harvested corresponds to the "irreducible minimum" residual amount. Only if this condition is met will the nitrate concentration be low, meaning that catch crops will not be useful or necessary for restoring the drainage water concentration to less than 50 $mgNO_3^-/l$.

Two other solutions could be assessed to reduce nitrate losses after grain maize:

1) sowing of less late maize varieties, enabling earlier harvest (mid to end of September), which would make it possible to sow a catch crop early enough for it to grow and be effective;
2) undersowing of a catch crop with maize. However, this technique still needs to be "optimized" since, first of all, there is no guarantee that the catch crop will grow sufficiently under the maize and, secondly, there is a risk that it will not be very effective if maize stover is "mulched" on the soil surface, significantly reducing the cover crop's access to light.

This undersowing technique was not evaluated in the simulation study due to a lack of data and specific parametrizing of the STICS model for this type of management technique. Specific studies and modelling would be required to be able to examine this question more closely.

6. **A mulch of chopped maize stover left on the surface of the soil does not reduce nitrate leaching after grain maize: it is preferable to bury the mulch, even though this technique has a limited efficacy**

Since catch crops are not effective for reducing leaching after grain maize (late harvest), a possible alternative may be the management of maize residues (stover) to reduce the nitrate concentration of drainage water. However, the simulations show that a maize stover mulch left on the surface of the soil has no effect on leaching, whereas buried residues trigger net immobilization of nitrogen in the soil, leading to a slight reduction in leaching (by 5 kg N/ha) and in the nitrate concentration. In this case, the nitrate concentration is only reduced by between 5 and 10 % (Fig. 3.16).

Consequently, for the purposes of reducing nitrate leaching, it is preferable to incorporate chopped grain maize stover immediately after harvest rather than to leave it as a mulch on the soil surface, even if the effect in terms of reduction in nitrate concentration is modest. In contrast, mulching is favourable in terms of carbon sequestration in the soil, because the decomposition of residues is slower and humification is greater.

7. **Oilseed rape and winter wheat volunteers are effective for reducing the nitrate concentration in drainage water, provided that the plant cover is dense and uniform**

The study results indicate that volunteers are effective for reducing leaching and the nitrate concentration in drainage water, provided that the plant cover is sufficiently dense and uniform (Fig. 3.17). In these ground cover conditions, oilseed rape volunteers prove to be almost as effective as a mustard catch crop and wheat volunteers almost as effective as an Italian ryegrass catch crop.

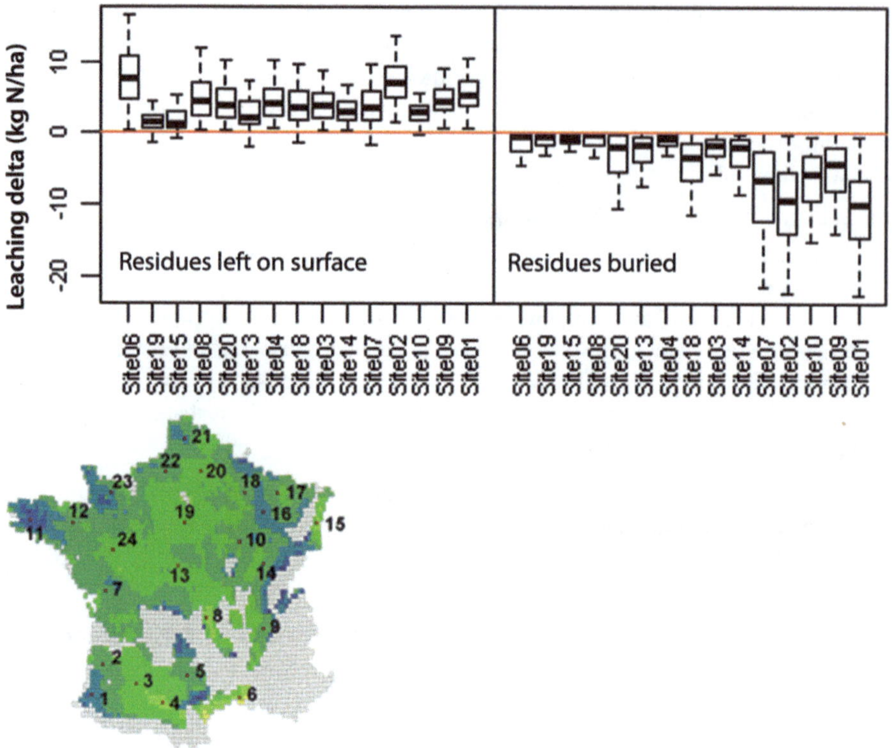

Fig. 3.16 Leaching differential (or delta) with residues *left* on the surface or buried, in comparison to the exporting of harvest residues from the previous crop, in the case of grain maize stover

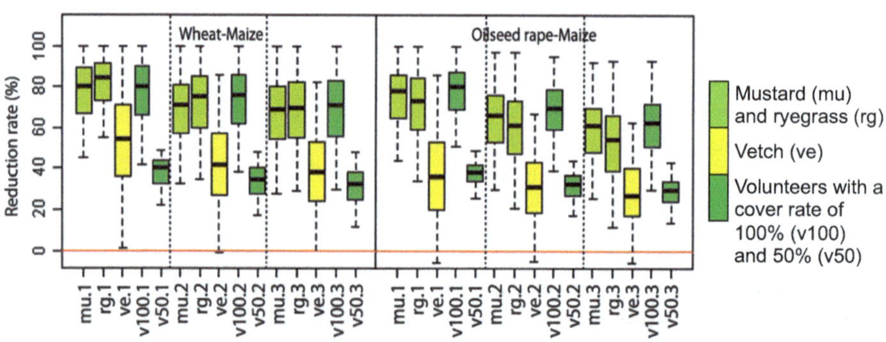

Fig. 3.17 Effectiveness of volunteers compared to sown catch crops. Nitrate concentration reduction rate as a function of different cover methods, initial mineral nitrogen status of the soil (1, 2, 3) and crop rotation. [Simulations on all sites; for emergence dates at 25/7 and 10/8, and destruction dates at 1/10 and 20/10 for the oilseed rape-wheat rotation and at 10/11 and 10/2 for the wheat-maize rotation]

This conclusion confirms the results of trials on oilseed rape volunteers in various soil and climate conditions in France. The simulations demonstrate that wheat volunteers are also effective when it comes to reducing leaching if they are well distributed spatially, with a high level of cover. The results indicate that cereals that emerge relatively early (early August, with a low dormancy) have sufficient growth and nitrogen uptake capacities to be able to trap nitrates before the drainage period. However, when the spatial distribution is uneven (in "strips" under the passage of the combine harvester, alternating with bare soil), volunteers are, on average, half as effective at reducing nitrate concentrations as when they are uniform. In the case of sparse and unevenly distributed volunteers, while the resulting cover crop is useful, it is not fully effective in terms of reducing nitrate leaching.

The ground cover upon which the effectiveness of volunteers depends is determined by the spatial distribution of seed losses at harvest. This distribution is dependent on the type of combine harvester used and whether it is equipped with a straw chopper and spreader enabling distribution of the seeds over the entire width of the cutter bar. The method used to promote the emergence of volunteers must therefore be appropriate to obtain sufficiently dense, uniform cover.

Stubble disking after harvesting oilseed rape and, more generally, the use of the stale seedbed technique, are often effective methods to encourage the germination of seeds that have fallen onto the soil and then to destroy plants that have emerged, preventing oilseed rape from becoming a weed in the next crop. In this case, the "nitrate-trapping" function of the vegetation cover might be reduced if the volunteers are destroyed too soon after their emergence, or if a second wave of oilseed rape emergence does not occur following the preparation of an initial stale seedbed. The effectiveness of oilseed rape or wheat volunteers for reducing nitrate leaching is not guaranteed, therefore, if the cover is not maintained for at least 2 months of growth.

8. **The number of potentially available days for the destruction of cover crops in the autumn is not a limiting factor in carrying out this operation, except for clay soils (>37 % clay)**

The number of days potentially available for the mechanical destruction of the catch crop (or volunteers) in the autumn was calculated on the basis of the daily moisture content of the soil's surface layer (0–30 cm) simulated using the STICS model from10 September to 10 December, the pressure exerted by the machinery used for the operation, and a water content level below which the risks of soil compaction are avoided.

The number of days available (Fig. 3.18) is potentially high in sandy or loam soils and is therefore not a limiting factor for the mechanical destruction of catch crops in the autumn (November and December). However, the number of days potentially available in clay soil situations (>37 % clay) may be low (<35 % of the days available per decade in November and December), particularly as the destruction date approaches December. Hence in clay soils it is preferable to destroy the cover crop early (end of October) in order to be able to till the soil with acceptable conditions (sufficient bearing capacity to be able to work) and avoid compaction

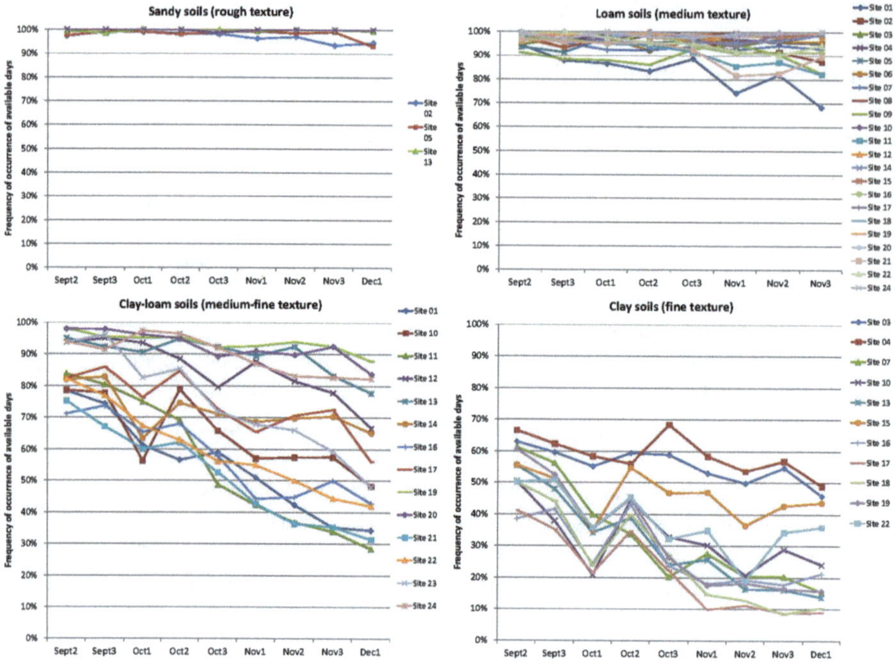

Fig. 3.18 Frequency over the 20-year period of days potentially available for each decade per site, for the main 4 soil texture types studied during the autumn

and a "clod-ridden" seedbed for the next crop. It should be stressed that this type of early destruction (mid to late October) is not incompatible with a significant reduction in leaching and nitrate concentrations in drainage water (see point 10 below).

It is worth noting that the study did not examine "heavy clay" soils (>45 % clay) due to their specific behaviour, which varies depending on the mineralogical clay type. They are difficult to take into account in the model because they present specific characteristics in terms of their physical behaviour (cracking, high level of plasticity, etc.) that current functional models do not adequately incorporate. In particular, for this type of high clay content soil, the feasibility of soil tillage operations is highly sensitive to hydric conditions. These high clay content soils are not actually widespread in French arable cropping systems, but they can pose genuine difficulties locally from time to time.

9. **The impact of Brassicaceae and Poaceae catch crops on the yield of the next main crop is slightly positive or nil, except occasionally in situations of short fallow periods when it can be negative. The impact is always positive with vetch.**

Overall, the simulations indicate positive effects of catch crops on the yield of succeeding maize crops, following wheat (long fallow period) and following silage maize (long fallow period, semi-late harvest) (Fig. 3.19). However, this yield

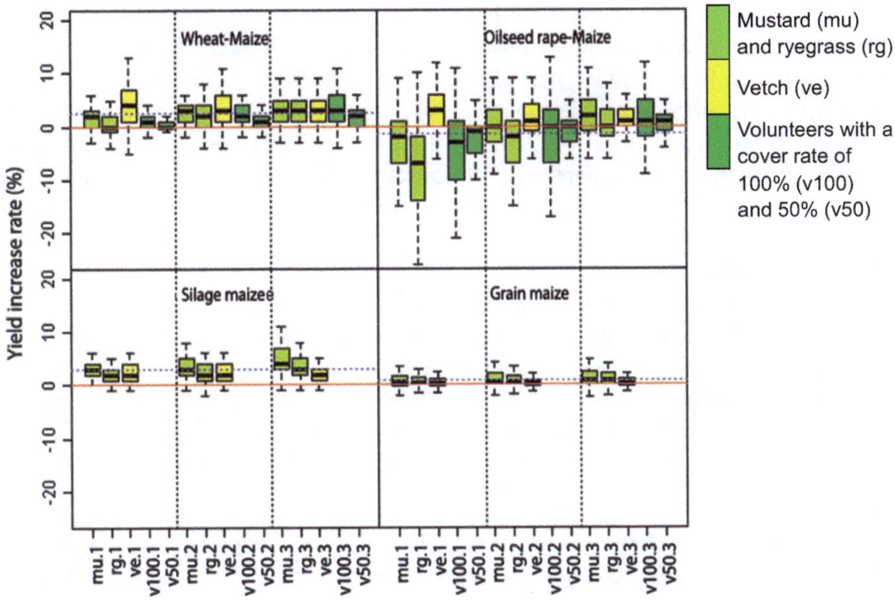

Fig. 3.19 Relative yield difference between a catch crop and bare soil (expressed as a function of reference bare soil yield) depending on the different cover methods, initial nitrogen status of the soil (1, 2, 3) and cropping rotation (*blue line* = average per rotation). [Simulations on all sites; for emergence dates at 25/7 and 10/8, and destruction dates at 1/10 and 20/10 for the oilseed rape-wheat rotation and at 10/11 and 10/2 for the wheat-maize rotation]

increase is moderate (with a median of 1–8 % depending on the climatic site). Furthermore, almost 75 % of the population of the simulations have a positive yield differential compared to "bare soil", indicating a globally positive effect of cover crops in terms of increasing nitrogen availability for the next maize crop. These simulations do not describe the long-term effects of catch crops on nitrogen (see Sect. 2.2.3.).

However, this effect on the yield of the next maize crop is almost nil in grain maize monocropping situations corresponding to late harvest of the previous crop, due to the low quantities of nitrogen taken up by catch crops that have emerged in late October.

In contrast, the simulations indicate highly variable effects in short fallow period situations after oilseed rape, with cases of lower wheat yields after catch crops in the sites with the lowest rainfall. More specifically, the median values for the simulated effects are often negative when the initial residual nitrogen content is low (irreducible minimum residual amount of 20 kgN/ha over 90 cm) for crops other than legumes (mustard, ryegrass and oilseed rape volunteers). The negative effect is potentially greater with ryegrass (median of −5 %, with values of up to −25 %). The lowering effects highlighted in simulations are even greater when the initial amount of mineral nitrogen in the soil is high and if the site is subject to high rainfall.

Finally, vetch almost systematically has a positive and highly significant median effect, in contrast with mustard and ryegrass (Fig. 3.19). These results indicate that in short fallow period situations, the phenomenon of preemptive competition for nitrogen (reduction in nitrogen availability) occurs in numerous soil and climate types and is particularly detrimental in the event of a poor initial mineral nitrogen status (in particular initial status N1). In these conditions, vetch, which "brings in" exogenous nitrogen by symbiotic fixation, substantially reduces this negative phenomenon and even leads to a positive effect in approximately 75 % of simulated cases.

To reduce the negative effects on the next crop in short fallow periods resulting from preemptive competition for nitrogen, one solution might be to sow a mixture of crops including a legume, or even a pure leguminous crop in the event of a low initial residual nitrogen content in the soil ("irreducible" initial status N1). Doing so would make it possible to benefit from the nitrogen obtained by symbiotic nitrogen fixation and reduce the C/N ratio of catch crop residues, thereby promoting nitrogen return to the next crop.

10. **In clay soils requiring autumn soil tillage, catch crops are always useful and are generally effective in terms of reducing nitrate leaching, even if they are destroyed early**

The question as to whether catch crops are effective is raised in early autumn destruction situations; in the study, the destruction dates of 1 and 20 October were considered to be early. These situations may correspond either to short fallow periods or to soil and climate conditions in which the soil needs to be tilled early before it is completely wet again, so that it can be worked in the correct conditions as recommended for clay soils (clay content > 37 %).

The simulations indicate that catch crops are generally effective for reducing leaching and nitrate concentrations in drainage water (Fig. 3.20) – although their effectiveness is not optimal – provided that cover crop emergence occurs before the end of August (simulated emergence on 25 August). In other words, catch crops are always useful and usually effective in short fallow period situations or when they have to be destroyed early (mid to late October). Consequently, it is always better to plant a catch crop – even if it has to be destroyed early in the autumn – than not to plant one at all if nitrate leaching has to be reduced (case of "Nitrate vulnerable zones").

In fact, the effectiveness of cover crops (volunteers or catch crops) for reducing nitrate concentrations in drainage water can be good (reduction rate always > 25 % and very often > 50 %), even if the maximum nitrate-trapping effect is not achieved. This result can be explained by the capacity of catch crops – and Brassicaceae in particular – to quickly take up mineral nitrogen from the soil (Fig. 3.20). It supports the use of a cover crop wherever possible during fallow periods, even in clay soil situations (37–45 % clay), and even more so in soils with a lower clay content.

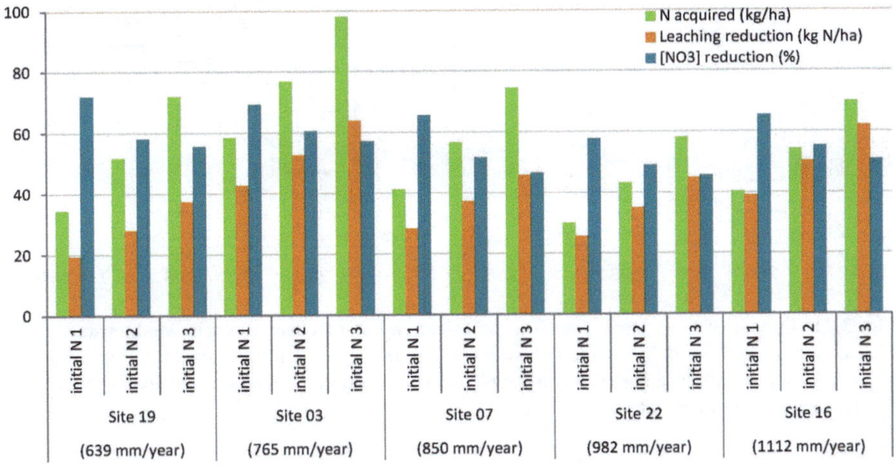

Fig. 3.20 Nitrogen acquired by mustard, reduction of leaching and nitrate concentrations in the event of early destruction (1/10) of catch crops having emerged between 25/7 and 25/8, for 5 contrasting sites (in terms of rainfall), on clay soils. (>37 % clay) and for the 3 initial mineral nitrogen levels (20–100 kgN/ha)

11. **In Mediterranean climates, emergence problems may occur but the interest of catch crops for reducing nitrate leaching is clear especially in years of excessive drainage.**

In Mediterranean climates, catch crops may be subject to high water stresses. This will limit germination or have a lethal effect on seedlings, thereby preventing the emergence of the catch crop. It is also possible that the high temperatures that can occur in this climate may prevent germination (>38 °C in seedbeds). Furthermore, even following their successful emergence, the plants may be subject to significant water stresses, inhibiting their growth in years when there is very little autumn rain.

Despite these development problems related to water shortages that occur in some years, the simulations indicate that catch crops are useful and, indeed, usually effective, for reducing leaching and nitrate concentrations in drainage water for a simulation scale of 20 years in Mediterranean climates, if the emergence date is optimal, as is indicated by the simulations for emergence on 10 September. In fact, "late" sowing (early September) leads to significant water stresses being "avoided". In addition, in contrast with early emergence dates (July and early August), which increase the risk of significant water stress and have a substantial impact on the nitrogen uptake capacity of catch crops, emergence dates in September are not incompatible with – and are, in contrast, preferable for – the effectiveness of catch crops (>50 % reduction in nitrate concentration in wet years).

Consequently, among all the sites, it is in Mediterranean climates that catch crops demonstrate the greatest effectiveness, when very late emergence dates are simulated and, especially, following grain maize (emergence on 25 October). This can be explained by the significantly higher temperatures and sunshine in the autumn and winter than at the other climatic sites, leading to significant growth of the plants during the autumn and even the winter. This makes the catch crops more

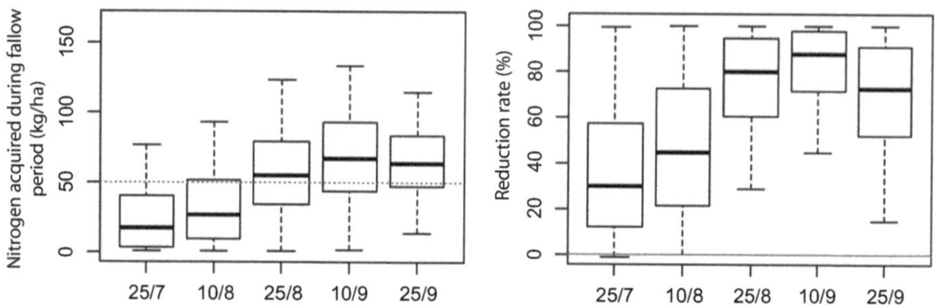

Fig. 3.21 Impact of emergence dates on the effectiveness of cover crops in Mediterranean climates. Nitrogen acquired during the fallow period (on the *left*) and nitrate concentration reduction rate compared to reference bare soil (on the *right*), as a function of emergence dates for mustard in a wheat-maize rotation at site 6

effective in terms of reducing nitrate leaching, especially in years with high rainfall when drainage occurs (Fig. 3.21).

By avoiding the high levels of water stress encountered at the end of the summer, sowing in early September increases the chances of success of a catch crop and their subsequent capacity to absorb mineral nitrogen from the soil. Moreover, to avoid negative effects on the next main crop, the simulations indicate that it is preferable to destroy the cover crop before 20 December in order to limit the effects of preemptive competition for water and nitrogen. This early destruction is a means of limiting the water consumption of catch crops, and hence the reduction of drainage and, ultimately, the impact on groundwater recharge.

Despite years without any drainage in this type of climate, the routine planting of catch crops as a preventive measure every year – especially in long fallow period situations – therefore emerges as an option to be favoured for reducing nitrate pollution over a 10-year period. Since long-range weather forecasts are not yet sufficiently reliable at present, it should be assumed that catch crops may prove to be of little use in years when drainage is nil or very low (approximately 1 year in two), bearing in mind that they are highly effective (nitrate concentration reduction rate in drainage water <75 %) in wet years (drainage > 100 mm).

12. **The application of livestock manure at the time of cover crop sowing is compatible with a reduction in nitrate leaching, as long as some essential rules are followed**

The application of slurry (pig slurry used as an example in the simulations) substantially increases nitrate concentrations in "bare soil" situations. It is important never to spread slurry in summer or autumn in the absence of plant cover – and hence even in the buried straw situation (which triggers net mineral nitrogen immobilization). Otherwise there is a risk of significantly increasing leaching and the nitrate concentration of drainage water.

It was decided to simulate the application of pig slurry since this application carries the highest risk in terms of increasing nitrate leaching due to the addition of organic waste products (OWP). Indeed, the nitrogen contained in this livestock slurry is mainly mineral (around 75 % in ammonium form). It is thus immediately available to plants but also for leaching. In addition, mineralization of its organic nitrogen is relatively rapid, further increasing the risks of leaching for applications in the summer and autumn.

Non-leguminous catch crops are effective for trapping and recycling nitrogen from the slurry and reduce nitrogen leaching compared to "bare soil" without the application of slurry in the majority of cases. However, this is only true if the amount of mineral nitrogen in the soil is low before slurry application and if the amount of slurry applied is not too high (75 kgN/ha). Mustard (Brassicaceae), which demonstrates faster growth and nitrogen absorption, is more effective than Italian ryegrass (Poaceae) in terms of reducing leaching and the nitrate concentration of drainage water following the application of slurry.

Although catch crops can reduce leaching and nitrate concentrations compared to "bare soil", they do not systematically restore drainage water concentrations to below 50 mgNO_3/l following the application of slurry, especially if the initial residual mineral nitrogen amount in the soil at the time of harvesting the wheat is above the minimum residual amount (initial status N1 with 20 kgN/ha over 90 cm) (Fig. 3.22).

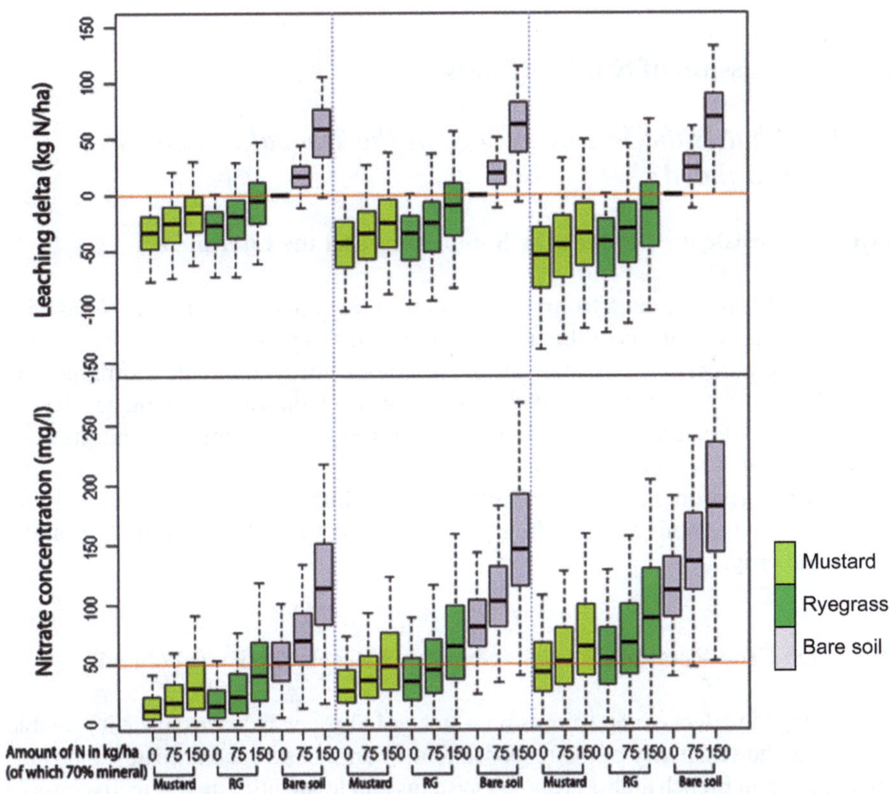

Fig. 3.22 Differential (or delta) for nitrate leaching (**a**) and nitrate concentration (**b**) by fallow period type, depending on the initial mineral nitrogen content of the soil (20, 60 and 100 kgN/ha) and the amount of nitrogen supplied by the slurry (0, 75 or 150 kg/ha)

Consequently, to avoid increasing nitrogen losses and the nitrate concentration of drainage water, the application of slurry combined with planting of catch crops can be conducted, provided that 4 conditions are met:

1. ensuring that the residual mineral nitrogen content of the soil at harvest is low (around 20 kgN/ha over 90 cm, it being imperative that it is less than 60 kgN/ha with little mineral nitrogen in the deep layers of the soil);
2. sowing a very fast-growing crop, such as a Brassicaceae (white mustard, for example);
3. implementing all necessary technical measures (sowing date and *ad hoc* method) to maximize the chances of successful emergence of the catch crop, i.e. a high emergence rate and acceptable spatial distribution of the plants, and non-exposure to high levels of water stress at the start of the establishment period. Otherwise, it is preferable to apply slurry only after emergence in order to verify that the cover crop is properly established and will be able to effectively absorb mineral nitrogen;
4. limiting the amount of slurry spread, so that the total nitrogen input does not exceed 75 kgN/ha.

3.4 Discussion of Study Results

3.4.1 What Is the Validity Range of the Typical Rotations Simulated?

Overall Consistency Between the Simulations and the Literature

Overall, the lessons drawn from this simulation study appear to be consistent with the results of the analysis of the literature in the different areas covered by the two approaches, namely: i) the order of magnitude of nitrogen quantities trapped by catch crops, ii) the effects of catch crops on water balance and drainage, iii) the effectiveness of catch crops and volunteers for reducing leaching and nitrate concentrations in drainage water and, conversely, iv) the poor efficacy of residue management (particularly surface mulches), and finally v) the factors explaining the effect of preemptive competition of catch crops for water and nitrogen on the next main crop.

To Which Crop Rotations Can the Simulated Results Be Extrapolated?

Simulating the effect of catch crops on water and nitrogen balances made it possible to analyse the sensitivity of the response to the variable soil and climate conditions encountered in French arable cropping systems and to quantify their effectiveness in comparison with other methods of fallow period management.

The three "typical" rotations selected represent, firstly, a significant surface area in French cropping systems and, secondly, correspond to the range of fallow period durations for which a catch crop can be planted, i.e.: i) an oilseed rape-wheat rotation, with a short fallow period, ii) a wheat-maize rotation, with a long fallow period, iii) a maize-maize rotation with a long fallow period and late harvest in the autumn.

The simulations revealed that the effectiveness of catch crops in terms of reducing the nitrate concentration of drainage water (reduction rate) is highly sensitive to the initial mineral nitrogen status of the soil and the soil and climate conditions. There is also an interaction with the duration of the fallow period and therefore the type of crop rotation. These factors determine nitrate leaching (amount of nitrate lost beneath the roots) and the nitrate concentration in drainage water.

The simulated results for the three "typical" rotations can be transposed to other non-simulated rotations, sometimes under certain conditions. These proposed relationships are detailed in Table 3.1.

The hypothesis is thus put forward that a number of rotations can be related to the simulated rotations, in terms of the effect of catch crops on water and nitrogen balance and the impact on the yield of the next crop. This is done for both short and long fallow period situations. Such a hypothesis is consistent with the results obtained from the analysis of the literature.

Conversely, rotations with a very short fallow period cannot be related to the short oilseed rape-wheat fallow period, since the time between the two crops is generally too short for adequate growth of the cover crop. These crop rotations into which a catch crop cannot reasonably be incorporated are: i) wheat-oilseed rape or barley-oilseed rape, ii) maize-winter wheat, iii) sunflower-wheat, iv) sorghum-wheat, v) soybean-wheat, vi) sugar beet-wheat, vii) potatoes-wheat.

3.4.2 Can Catch Crops Mitigate Nitrate Leaching Situations with Excess Mineral Nitrogen in the Soil at the Time of Harvest?

Catch crops unambiguously mitigate situations whereby residual soil nitrogen contents at harvest exceed the irreducible "minimum" when the level is not too high, as is the case for initial status N2 (corresponding to 60 kgN/ha over 90 cm) in short and long fallow period situations. But they are insufficiently effective in situations following grain maize.

Despite their high nitrogen trapping ability, non-leguminous catch crops, and legumes even more so, do not always reduce the nitrate in the drainage water to less than 50 mgNO_3/l in situations where there was a high initial quantity of mineral nitrogen in the soil, as simulated in the case of initial status N3 (corresponding to 100 kgN/ha). This status aims to simulate cases of incorrect nitrogen fertilizer management or cases of excessive application of organic waste products that are readily mineralizable or have a high mineral nitrogen content.

Table 3.1 Relating of other crop rotations to the simulated typical rotations

Fallow period	Short	Long, summer harvest	Long, early autumn harvest	Long, late autumn harvest
Simulated typical rotation	Oilseed rape – Wheat	Wheat – Maize	Silage maize – Maize	Grain maize – Grain maize
Other related rotations	Peas – Wheat	Wheat - Spring barley	Sugar beet (early harvest) – Spring peas	Sugar beet (late harvest) – Spring peas
	Field beans – Wheat	Wheat – Spring peas		
	Wheat – other winter cereal …	Wheat – Sugar beet	Sunflowers – Sorghum …	Soya – Maize
		Wheat – Potatoes		Soya – Maize
		Wheat (common or durum) – Sunflowera		
		Wheat – Soya		
		Wheat – Sorghum		
		Durum wheat – Melon …		
Comments	After a leguminous crop: *the "irreducible" residual mineral nitrogen content of the soil at harvest may be higher than the minimum amount (initial status N1 in the simulations) and could correspond more closely to an intermediate initial status (initial status N2 in the simulations)*	Wheat – Spring peas: *the simulated effects of cover crops on maize may not represent the potential effects on the leguminous crop (symbiotic nitrogen fixation may compensate for any lack of mineral nitrogen = preemptive effect for nitrogen)*	*The need to have a mineral nitrogen status at harvest of the previous crop that is as low as possible (low quantities in deep soil horizons, horizons 60–90 or > 90 cm) → Objective of nitrogen fertilizer application management: to be very carefully adjusted in order to have an initial status close to simulated status N1 (20 kgN/ha maximum over 90 cm)*	

It is therefore necessary to once again highlight the fact that the first requirement to control nitrate leaching is to adjust the amount of nitrogen fertilizer applied to main crops as carefully as possible. This nitrogen fertilizer management is also a prerequisite for the effectiveness of catch crops, particularly when the objective is to produce drainage water with a low nitrate content (and, where possible, below 50 mgNO_3/l). This should make it possible to avoid the occurrence of excessively high residual mineral nitrogen contents in the soil at the time the main crop is harvested and, in particular, a distribution at the "base" of the soil profile, not "retrievable" by catch crops or previous crop volunteers.

3.4.3 An Approach for Defining the "Optimal" Date of Catch Crop Emergence and Destruction by Multi-criteria Analysis of Simulation Results

This simulation study demonstrates how useful it is to use dynamic models for simulations that take into account processes occurring in interaction (or competition) with one another, with timely competition. It highlights the fact that it is necessary to adjust management techniques accordingly on the basis of soil and climate conditions. This conclusion is consistent with the analysis of the experimental results available in the literature. Both the results of the simulation study and those contained in the literature demonstrate that the impact of catch crops on drainage, nitrate leaching and the next crop depend on numerous parameters related to the catch crop itself (amount of nitrogen uptake, C/N ratio). The effectiveness of catch crops therefore depends, concomitantly, on: i) the type of crop (non-leguminous versus leguminous), ii) the sowing date (and hence the emergence date), iii) the date of destruction and incorporation, but also iv) the environment, and hence the climate and soil type (texture and water holding capacity).

"Optimal" management techniques can therefore be defined subsequently on the basis of the simulation results, for the wheat-maize rotation, for example, and for each soil and climatic site studied. For each site, each of the three catch crop types and each of the three initial mineral nitrogen levels, the method consists in defining the emergence and destruction date pairing that leads to the best overall performance over a simulation period of 20 years through the successive application of several criteria. The criteria selected, their values and their hierarchization are as follows:

1. a high efficacy of catch crops for reducing nitrate leaching, characterized by an interannual nitrate concentration average of less than 50 $mgNO_3/l$ (or at least < 60 $mgNO_3/l$ if this condition is not met) <u>and</u> a concentration reduction rate of over 75 % of the rate obtained for the most effective emergence/destruction date pairing;
2. the absence of any negative impact on the next crop, i.e. a yield for the next crop greater than or equal to that obtained without the use of a catch crop;
3. a reliability of catch crop efficacy, defined as a low probability (less than 15 % or, at the minimum probability of +5 %) of obtaining a reduction in nitrate concentration of less than 50 %;
4. the lowest possible reduction in drainage (to within 5 % of the minimum) so as to limit the impact on groundwater recharge.

Each criterion is applied according to the hierarchy defined above: the test to sort the "emergence/destruction" date pairings is only applied for the data combinations that have already met the previous criteria (Fig. 3.23).

It must be stressed that the results obtained and presented here are highly dependent on the criteria selected. They should be seen as an illustration of the process; hence other results could have been obtained, either by altering the hierarchy of the

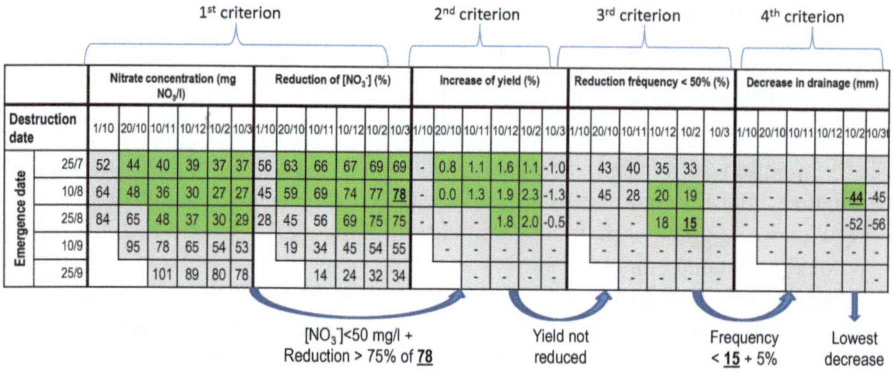

	Nitrate concentration (mg NO3/l)						Reduction of [NO3-] (%)						Increase of yield (%)						Reduction fréquency < 50% (%)						Decrease in drainage (mm)					
Destruction date / Emergence date	1/10	20/10	10/11	10/12	10/2	10/3	1/10	20/10	10/11	10/12	10/2	10/3	1/10	20/10	10/11	10/12	10/2	10/3	1/10	20/10	10/11	10/12	10/2	10/3	1/10	20/10	10/11	10/12	10/2	10/3
25/7	52	44	40	39	37	37	56	63	66	67	69	69	-	0.8	1.1	1.6	1.1	-1.0	-	43	40	35	33	-	-	-	-	-	-	-
10/8	64	48	36	30	27	27	45	59	69	74	77	**78**	-	0.0	1.3	1.9	2.3	-1.3	-	45	28	20	19	-	-	-	-	-	**-44**	-45
25/8	84	65	48	37	30	29	28	45	56	69	75	75	-	-	-	1.8	2.0	-0.5	-	-	-	18	**15**	-	-	-	-	-	-52	-56
10/9		95	78	65	54	53		19	34	45	54	55		-	-	-	-	-		-	-	-	-	-		-	-	-	-	-
25/9			101	89	80	78			14	24	32	34			-	-	-	-			-	-	-	-						-

Fig. 3.23 Example of application of the multi-criteria analysis to define the optimal emergence and destruction dates according to the defined criteria (the data combinations selected at each stage and hence possible are shown in *green*)

criteria, on the basis of priorities depending on the soil and climate conditions considered, or by modifying the value of the criteria.

The results are presented by site, crop type and initial mineral nitrogen status of the soil (Fig. 3.24), because these three factors are relevant when it comes to explaining the variability of the simulated results. The key lessons drawn from the "multi-criteria optimization" are as follows:

- As expected, the optimal emergence and destruction date pairings for the cover crop depend on the climatic site. They are distributed according to a dual North-south and West-east axis in France.
- However, the optimal "emergence/destruction" date pairings depend on the initial mineral nitrogen status of the soil at the time of harvesting the previous crop.
- Mustard and ryegrass meet all the criteria for almost all the sites (with the exception of Mediterranean site 6, with initial status N3), irrespective of the initial mineral nitrogen status when the previous crop was harvested.
- Vetch (a leguminous crop) does not meet the criteria for the N2 and N3 initial statuses at the time of the previous crop harvest (60 and 100 kgN/ha respectively) in a large proportion of the sites, particularly in the Southern half of France, due to a lack of effectiveness in terms of reducing the nitrate concentration of drainage water. However, the legume meets the multi-criteria selection if the initial nitrogen status is N1 at the time of harvest of the previous crop (20 kgN/ha) for all the sites. Leguminous crops used as "pure" catch crops are useful for reducing nitrate leaching, but if the aim is to reduce the nitrate content of drainage water to below 50 $mgNO_3/l$, it is necessary to make sure that the residual mineral nitrogen content of the soil is low (close to N1 status) before sowing a legume.
- The optimal emergence dates are selected between 10 and 25 August, with a few exceptions on 10 September, for the two non-leguminous crops (mustard and ryegrass): this indicates that to be effective the catch crops do not need to be sown too soon after wheat harvesting. However, the optimal emergence date is earlier for vetch (case of initial status N1), from 25 July to 10 August.

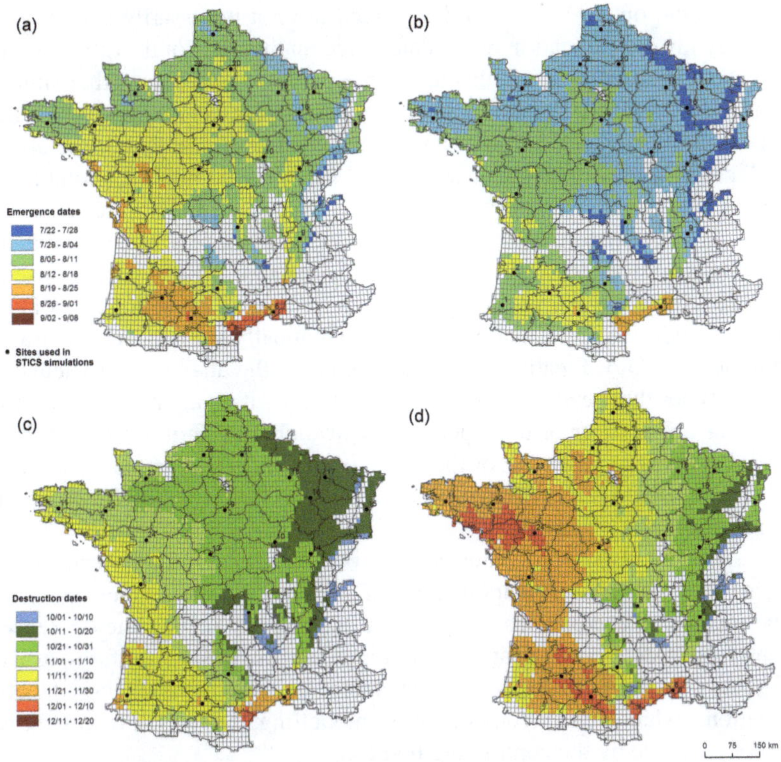

Fig. 3.24 Optimal emergence and destruction dates of the catch crops per climatic site (1–24), depending on initial mineral nitrogen status (N1, N2, N3) and crop type (mustard, ryegrass and vetch), determined by multi-criteria analysis (considering, simultaneously, the impact on nitrate concentrations, the yield of the next crop and drainage) on the basis of the simulation results. "<60 $mgNO_3/l$" indicates that the value obtained is below a nitrate concentration of 60 $mgNO_3/l$ (with the <50 $mgNO_3/l$ criterion not having been met)

- Contrary to the generally accepted idea that it is preferable to destroy a catch crop as late as possible, the optimal destruction dates are generally early, falling within the period between 20 October and 10 December. This is also true for oceanic climate zones. Hence, other than in exceptional cases, the optimal period for destruction is the autumn. However, the higher the initial mineral nitrogen content of the soil, the later the date is. This effect is particularly marked for ryegrass, for which the optimal destruction dates are frequently 10 February with an initial status of N3.

The optimal dates determined in this way represent an initial evaluation and illustrate the process that could be implemented for each soil and climate situation in France to optimize catch crop management techniques. The results obtained are contingent upon the simulated emergence and destruction dates: simulations performed with shorter time-frames would make it possible to specify the optimal dates. Hence the optimal dates so determined are not necessarily true optimums. They merely give an indication of the date range on the basis of the different French soil and climate conditions simulated, in order to have catch crops that simultaneously meet the various criteria in terms of efficacy or avoidance of adverse effects.

An alternative hierarchization of the criteria for the assessment of the effects of the catch crops taken into account would lead to different "optimal" combinations. Thus, if the objective had simply been to obtain a reduction in nitrate leaching, the "optimal" destruction dates would have been later than those obtained in this multi-criteria analysis. However, it should be noted that the sorting criterion used was already strict since it was based on a nitrate concentration reduction rate of at least 75 % of the maximum level, which almost systematically leads to a concentration of less than 50 $mgNO_3/l$. Finally, in order to complete the analysis – particularly for certain soil types that need to be tilled in the autumn – it would be necessary to take into account the number of days potentially available to optimize the destruction date. All this gives a clear view of the scope of the optimization work that remains to be done in order to have regional-based "references" adapted to the different soil types and cropping systems.

With a view to improving water quality (both underground and surface) on the scale of a watershed or of the capture zone of an alluvial aquifer, other spatial scales must be taken into account when optimizing catch crop management. Hence, the location of catch crops each year, the cropping plan and the type of farm, as well as hydro-geological properties, must be considered. It should be remembered that such investigations, which demand other types of modelling, did not fall within the scope of the request made by the contracting bodies.

3.4.4 Cover Crops: What Other Ecosystem Services Can They Provide?

Given the results of the simulations and the analysis of the literature, it appears to be entirely possible that cover crops used as "nitrate-trapping" catch crops may provide several ecosystem services simultaneously:

1. a "green manure" effect to reduce nitrogen fertilizer application or increase the yield (or protein content) of the next crop;
2. sequestration of organic carbon and nitrogen in the soil and better nitrogen mineralization in the long term with cumulative use over time;
3. reduction of certain pests and diseases;
4. protection of the soil against erosion thanks to ground cover before destruction and to catch crop residues if they are left as a mulch on the soil surface after destruction.

However, it is important to highlight the possibility of incompatibilities between the ecosystem services expected from cover crops.

Cover Crops and Soil Protection

If the soil needs to be tilled early in the autumn (case of clay soils, for example), the water erosion reducing function will be limited and the cover crop – which must be destroyed early – will not provide an optimal nitrate-trapping effect, although it will generally be useful for reducing nitrate concentrations in drainage water.

Conversely, in loam soils with poor structural stability, a late destruction date (early spring, just before sowing the next main crop) could make it possible to simultaneously maximize the "nitrate-trapping" effect and the protective effect against soil surface sealing; consequently, the later the catch crop is destroyed and the closer this date is to the planting of the next main crop, the greater this positive dual effect will be. But there is a risk that the drainage-reducing effect will be increased and that greater preemptive competition for nitrogen will occur, with a potentially negative effect on the next crop, if nitrogen fertilizer application is not adjusted (dose, division into several applications and application dates).

Cover Crops and Stale Seedbeds

Another question is raised concerning the "compatibility" between the practice of using stale seedbeds to manage weeds (widely used in organic agriculture or regularly implemented in certain cropping systems) and the "nitrate-trapping" function. This weed control practice generally consists in tilling the soil just after the previous crop is harvested to encourage the emergence of a proportion of the weed seed stock and previous crop volunteers (which may become weeds). This technique does not appear to be incompatible with the planting of catch crops – except in specific cases – for the following reasons:

- The stale seedbed will allow weeds to germinate and possibly emerge. The effectiveness of stale seedbeds begins whenever weeds have germinated, meaning that weeds do not necessarily need to have emerged before the soil seed stock is reduced by tilling the soil. This can be explained by the fact that germination is an irreversible phenomenon and that germinated seeds are highly sensitive to water stresses and/or mechanical tillage of the soil, which generally destroy the seedling.
- Delaying the sowing of a catch crop by 2–3 weeks to enable the practice of stale seedbed preparation may make it possible to achieve a good level of effectiveness for the catch crop in terms of reducing nitrate leaching if emergence occurs in late August to early September at the latest, depending on the climatic site.
- Herbicides are not generally used on catch crops for weed control. As a result sowing these crops can allow the emergence of weeds, meaning that catch crops can, in themselves, play the role of stale seedbeds. In this case, a fast-growing Brassicaceae – competitive in terms of the environmental resources – could be sown to compete with the weeds and reduce their growth.

- The destruction of the catch crop in the autumn should also enable the destruction of weeds before they produce new seeds, provided that the catch crop is uniformly distributed to stimulate competition against the development of weeds and that the weed flora in the field is not made up of annual types with very short cycles capable of producing seeds less than 2 months after their emergence.

In conclusion, it appears that the dual practice of stale seedbeds and catch crop planting requires more precise and probably more complex fallow period management.

Cover Crops and Pests

Since catch crops can promote certain pests (slugs, soil-borne diseases, etc.), the plant health risk must be taken into account when selecting the type of cover crop and possibly even the variety.

Although previous crop volunteers (oilseed rape or cereal) can be effective for reducing nitrate leaching, they may lead to the development of pests and diseases as potential host plants for pests that attack the main crops in the rotation. Consequently, it is important to analyse the risk of triggering a multiplication of pests and diseases as a result of their use. This risk will depend on the level of infestation of the previous main crop, the biological properties of the pest considered and, especially, the rotation "return time" (i.e. the length of time before the crop is planted again) for the main crop (short rotation versus long rotation).

For maize stover management, the simulation study indicates that it is preferable to bury it to obtain a reduction – albeit modest – in the nitrate concentration of drainage water, rather than to leave it as mulch on the soil surface. While incorporation can reduce the protective effects of soil mulched with chopped maize stover against water erosion, mechanical destruction (ploughing, for example) is preferable on order to control corn borer and corn root worm and to reduce fusarium head blight in wheat in maize-winter wheat rotations. Chopping the maize stover and incorporating it into the soil by deep tillage therefore fulfils several functions simultaneously.

Cover Crops and Pesticide Transfer

Recent results have demonstrated that catch crops can also have a positive effect in terms of reducing pesticide transfer – and especially herbicide transfer – in irrigated maize monocropping situations. This effect can be explained by the modification of the hydrodynamic properties of soils and hence the drainage and leaching of pesticides with similar dynamics to nitrate leaching. It is also due to the uptake and degradation of pesticides by the soil's microflora, the development of which is probably stimulated by the incorporation of catch crop residues. These potential effects – which need to be confirmed by additional studies – would further increase the value of catch crops in terms of improving the quality of drainage water returned to groundwater reservoirs.

Cover Crops on Sloping Land

On sloping land, the crucial question may be how to reconcile the erosion-reducing effects with nitrate leaching reducing effects, without reducing the availability of water in the soil for the next crop.

To be effective against erosion and runoff, catch crops need to present adequate soil cover density and uniformity (efficacy threshold for a cover level > 30 %) and be fast-growing, and they also need to stay in place for a sufficient period of time, including in the spring. The same conditions govern their effectiveness against nitrate leaching.

However, prolonging the growth period of a catch crop increases the amount of water it uses and therefore increases the risk of preemptive competition for water for the main crop in the spring, particularly given that the catch crop will begin to grow again at this point. So favouring the "anti-erosion" function could reduce groundwater recharge unless, by significantly reducing runoff (sloping land), the catch crop increases water infiltration and hence soil water recharge and drainage compared to bare soil. This effect may be more evident in no-tillage situations.

Few data are available in the literature concerning this specific problem. It might be a good idea to analyse the combined effects and determine the conditions that would be favourable, or otherwise, for the joint attainment of all three objectives by conducting simulation studies using a validated model integrating run-off and vertical transfers, or even subsurface flows, on sloping land.

3.4.5 How Can Catch Crops Be Adapted Locally to Allow Their Broadest Possible Integration into Current Cropping Systems?

Potentially, catch crops could have a key role to play for redesigning cropping systems, based on the various potential ecosystem services they may provide. For this, their management techniques need to be adapted on a regional (or even local) basis to increase their positive effects and reduce their potential negative effects on the next crop. There is no doubt that adapting management techniques to local conditions will be crucial to the successful integration of catch crops within French cropping systems in order to avoid negative effects on the next crop and obtain the ecosystem services targeted. The choice of crop type, catch crop sowing and destruction dates and residue management method can be determining factors in whether or not the desired functions are obtained. Consequently, no single management technique is likely to be theoretically optimal for all local conditions, particularly if the objective is obtain several functions and services simultaneously. This study demonstrates that it is necessary – and this is also true for a main crop – to adapt catch crop management methods to the cropping rotation and local soil and climate conditions.

Furthermore, management techniques must be compatible with the organization of labour for the farm as a whole. However, the planting and destruction of catch crops may conflict with other farming operations and, in particular, summer and autumn harvests, autumn sowing and livestock-related activities. These various operations take priority and fallow period management methods need to be carefully defined in order to fit as seamlessly as possible into the farm's work schedule thereby maximizing the development of catch crop use. The approaches developed by agronomic research and agricultural development to aid agricultural equipment decision-making processes could be redeployed for this purpose.

Finally, it is important not to neglect the technical aspects of sowing in no-tillage or minimum tillage situations, in which appropriate methods must be implemented to increase the chances of success of cover crop planting.

While volunteers can be effective for reducing the nitrate concentration of drainage water, they may pose potential problems by favouring certain pests and diseases. It may therefore be preferable – or even necessary – to sow species belonging to different botanical families in order to produce complementary ecosystem services ("nitrate input or green manure effect", reduction of certain pests such as beet nematodes, prevention of the spread of soil-borne diseases, etc.). The use of a diversity of crop types in rotations should be favoured. Management of ground cover also makes it preferable to use specifically sown catch crops, for which the sowing amount and date can be adjusted on the basis of soil type and soil tillage management options practised by the farmer.

Finally, it is important to highlight the fact that, in the medium and long terms (>10 years), the routine practice of using catch crops in cropping systems during each available fallow period leads to an increase in carbon and nitrogen sequestration in the soil. Ultimately, therefore, it leads to an increase in the amount of nitrogen supplied to the soil. This routine practice does not lead to a decrease in the effectiveness of catch crops for reducing the nitrate concentration of drainage water insofar as the integration of catch crops constitutes a specific practice in its own right within the cropping system. However, it must be stressed that nitrogen fertilizer application must be reduced to obtain the same potential yield for the next cash crop, since organic nitrogen sequestration leads to a significant increase in net annual nitrogen mineralization of the soil.

References

Brisson N, Mary B, Ripoche D, Jeuffroy M-H, Ruguey F, Nicoullaud B, Gate P, Devienne-Barret F, Antonioletti R, Dürr C, Richard G, Beaudoin N, Recous S, Tayot X, Plenet D, Cellier P, Machet J-M, Meynard J-M, Delécolle R (1998) STICS: a generic model for the simulation of crops and their water and nitrogen balances. Theory and parametrization applied to wheat and corn. Agronomie 18:311–346

Brisson N, Launay M, Mary B, Beaudoin N (2008) Conceptual basis, formalisations and parameterisation of the STICS crop model. Éditions QUAE, INRA, 78026 Versailles cedex, p 304

Dorsainvil F (2002) Évaluation par modélisation de l'impact environnemental des modes de conduite des cultures intermédiaires sur les bilans d'eau et d'azote dans les systèmes de culture. Thèse docteur-ingénieur Ina-PG., p 124

Dorsainvil F, Dürr C, Justes E, Carrera A (2005) Characterisation and modelling of white mustard (Sinapisalba L.) emergence under several sowing conditions. Eur J Agron 23:146–158

Dürr C, Aubertot JN, Richard G, Dubrulle P, Duval Y, Boiffin J (2001) SIMPLE: a model for SIMulation of Plant Emergence predicting the effects of soil tillage and sowing operations. Soil Sci Soc Am J 65:414–423

Chapter 4
Conclusions and Outlook

Eric Justes

4.1 The Contribution of the Study: A Complement to the Analysis of the Literature to Help Quantify the Effects of Catch Crops and Fallow Period Management Methods in French Soil and Climate Conditions

The analysis of the literature demonstrated the indisputable effectiveness of catch crops for a broad range of soil and climate conditions where there is significant drainage. The main factors determining drainage and nitrate leaching processes, as well as those explaining how catch crops are necessary and effective for reducing the nitrate concentration of drainage water, have been demonstrated and illustrated. However, the effectiveness of catch crops varies depending on the bibliographic references, although this variability has not been clearly structured or modelled. Indeed, few extension studies using statistical or functional modelling have been published, whereas numerous experimental studies in specific conditions are available.

The simulation study thus made it possible to quantify the effects of catch crops and other fallow period management methods on the basis of several factors, on a local scale and over a period ranging from harvest of the crop preceding the catch crop to harvest of the subsequent crop. The variability of these effects was analysed

The original version of this chapter was revised.
An erratum to this chapter can be found at https://doi.org/10.1007/978-94-024-0986-4_5

E. Justes (✉)
UMR AGIR (AGroécologie, Innovations & teRritoires) 1248, University of Toulouse, INRA, INPT, Centre de Recherches INRA Occitanie-Toulouse, 24 Chemin de Borde Rouge, Auzeville, CS 52627, Castanet Tolosan 31326, France

UMR SYSTEM Univ Montpellier, Cirad, Inra, Montpellier SupAgro, CIHEAM-IAMM, 2, place Pierre Viala, Bat. 27, Montpellier 34060 Montpellier cedex 2, France
e-mail: eric.justes@inra.fr; eric.justes@cirad.fr

E. Justes (ed.), *Cover Crops for Sustainable Farming*,
DOI 10.1007/978-94-024-0986-4_4

as a function of climate, incorporating soil types (3–4 soil types per climatic site) and the initial mineral nitrogen status of the soil at the time the previous crop was harvested (3 very different levels). The effectiveness of catch crops was therefore determined in a broad variety of French soil and climate types and regions. For some of these, few experimental data were available. The climatic variability, tested for the 24 sites selected, was evaluated on the basis of the current climate represented by the last 20 years (1988–2008). This made it possible to determine the variance associated with the effectiveness of catch crops and other fallow period management methods for a broad range of climatic conditions, taking into account "exceptionally" wet or dry years.

Ultimately, the simulations indicate that, in the context of France and on a scale of 20 climatic years, catch crops are effective in terms of reducing the nitrate concentration of drainage water. This is true for both short fallow periods and long fallow periods after a summer harvest. More specifically, the simulations indicate that catch crops are useful – or even often necessary – to obtain nitrate concentrations below 50 mgNO_3/l in drainage water, even in situations of judicious and optimal nitrogen fertilizer application. This can be explained by the fact that although management of nitrogen fertilizer application is a prerequisite, it is not in itself sufficient to obtain drainage water with a low nitrate concentration. In fact, it is summer and autumn mineralization (warm and sufficiently moist soils) that determines the risks of leaching to a greater extent than the residual amount present at the time of harvesting, when the latter is low (corresponding to the irreducible minimum residual mineral nitrogen content). In this case the application of inorganic nitrogen fertilizer to the catch crop would simply increase the leachable nitrate stocks and cannot therefore be justified.

However, the simulations show that catch crops are not effective after a late autumn harvest (following grain maize), indicating the need for very strict and careful management of nitrogen fertilizer application to limit nitrate leaching. More generally, the simulations indicate that in situations where there was a high initial quantity of mineral nitrogen in the soil, cover crops do not systematically restore drainage water nitrate concentrations to below the regulatory level.

The study also determined the degree of effectiveness of catch crops on the basis of crop type used (Brassicaceae, Poaceae or leguminous crop) and their management method (sowing and destruction date). Hence a multi-criteria analysis led to "optimal" emergence/destruction date pairings being proposed, for each climatic site and crop type. This study illustrates, for the first time, a broader approach that could be applied to optimize catch crop management techniques on the basis of soil and climate type, initial residual mineral nitrogen content in the soil and the next main crop, or even other factors.

The study yielded some particularly original results for two aspects concerning catch crop management. First of all, the simulation study of the "sowing-emergence" phase yields new results, whereas very few modelling studies are available in the literature regarding this phase of the planting process, despite the fact that it is so crucial for summer sowing in temperate and Mediterranean conditions. The simulations performed indicate that catch crops can be successfully planted in a broad range of situations, as long as the sowing date is adapted to suit the soil and climate conditions. Furthermore, the analysis of the number of days potentially available for destruction of catch crops is also innovative; although studies have been conducted for different cropping systems and for technical soil tillage or harvesting operations, no scientific articles were found examining the issue of catch crop destruction. The simulations therefore indicate that the number of days potentially available for the destruction of cover crops in the autumn is not a limiting factor in carrying out this operation, except in clay soils.

Management techniques should thus be adapted to the local soil and climate conditions and to the cropping system in order to simultaneously maximize the effectiveness of "nitrogen trapping" and minimize any adverse effects on the next crop. This adaptation of management techniques – particularly the sowing and destruction dates – should also take into account the other "services" that might be provided by catch crops, depending on the local context (preventing erosion, restoring water resources, etc.). On a broader scale, fallow period management is one of the components involved in (re)designing cropping systems that are more environmentally friendly and attribute a value to ecosystem services.

4.2 Remaining Crucial Research Questions to be Explored

Research Questions to be Explored

This study has highlighted the need to obtain more data on cover crops, as well as the requirement to develop a broad coordinated research programme examining this systemic issue in order to advance knowledge in the field of cover crop effectiveness.

1. **Contribution to main crop pest and disease management**. There is no doubt that research work needs to focus on this theme if catch crops are to be used extensively in France, since it is essential to ensure that neither their favourable side effects nor their adverse ones are overlooked. This raises several issues:
   - the compatibility of the stale seedbed technique with catch crop planting, which could benefit from simulation studies investigating effects on weed flora using a FLORSYS-type model;
   - the potential plant health problems that may arise as a result of the use of volunteers (oilseed rape or cereal) as cover crops; epidemiological studies conducted over the time-scale of one rotation should make it possible to analyse the risks on the basis of cropping system and soil tillage method;
   - the impact of catch crops on certain particularly problematic pests, which will need to be the focus of research or R&D studies in order to anticipate any risks related to the emergence of new parasite problems.

(continued)

2. **Simultaneous maximizing of "nitrate-trapping" effects (catch crop) and "green manure" effects** to reduce the use of mineral nitrogen fertilizers, the manufacture of which requires large quantities of fossil energy.

   This issue relates to the use of leguminous plants as catch crops and raises the question of the efficacy of nitrate trapping and symbiotic nitrogen fixation depending on the mineral nitrogen status of soils. It is particularly important that we examine this question if we want to develop the use of mixtures combining one or more non-leguminous and leguminous crops, in order to simultaneously obtain the two functions: "nitrate trapping" and "green manure".

   Research studies investigating crop combinations are required in order to determine effective mixtures. Although advances have been made in this area as a result of agronomic expertise – particularly in America, but also in Europe – the analysis of the literature revealed a lack of methodological studies. This represents an obstacle to the practical design of these types of mixtures, and their adaptation to the various soil and climate types found in France. It also makes it difficult to model plurispecific cover crops and perform simulation studies examining their effects on water and nitrogen balances, and hence on the reduction of nitrate leaching. The issue of the composition of crop mixtures (and the associated techniques) is also raised for plurispecific non-leguminous cover crops, aimed at maximizing nitrate trapping, as well as tolerance to drought or biofumigation.
3. **The choice of cover crop types and varieties and their genetic improvement**. Apart from certain types of Brassicaceae – seed companies have invested significant resources in Europe and France in these – it should be highlighted that investments in genetic research and plant improvement needs to be encouraged, particularly with a view to identifying and characterising new genetic resources vis-à-vis the management of pests, but also for other agro-ecological functions. It is possible to hypothesize that crop types currently little used in France for agricultural production may join the range of solutions available to farmers for the purposes of controlling certain pests. This is probably an area that could lead to genetic innovations, ultimately making it possible to reduce pesticide use.

   The study also reveals the lack of ecophysiological knowledge relative to the plant types currently used as catch crops, especially those that are not grown to produce grain or fodder.
4. **The management of cover crops in no-tillage systems**. Cover crops are widely used to compensate for the absence of soil tillage (since they favour water infiltration, decompaction, etc.) in no-tillage contexts, which modify the conditions for destruction and management of cover crop residues. However, the scientific literature does not enable any conclusions to

(continued)

be drawn in a number of areas, notably: the mechanical destruction methods that can be use in regions in which lethal frosts are uncommon in order to prevent the need to use systemic herbicides to destroy them; the effects of mulching (compared to burial) on the mineralization of cover crop residues, losses due to volatilization of ammonia, denitrification and N_2O emissions, or pests.

5. **Planting of catch crops by undersowing**. The difficulty related to incorporating catch crops into grain maize monocropping systems highlights the need to increase the effectiveness of undersown catch crops. Two priority questions need to be analysed: (1) Is it possible to reconcile the use of high-potential but earlier maize varieties with undersowing? (2) Which types of crop (and methods) are suitable for undersowing (tolerance to shade)?
6. **The link between catch crops and climate change**. The indisputable rise in temperature and CO_2 concentrations, as well as the seasonal variability of rainfall amounts and the risk of extreme events – especially autumn droughts – are liable to have an impact on the "nitrate-trapping" effectiveness of catch crops. Climatic changes may also alter their impact on drainage and hence potentially on groundwater recharge. These effects could be estimated by simulations using future climate scenarios.

   Moreover, the evaluation of the contribution of catch crops to the greenhouse gas balance of the agricultural sector is likely to be modified as a function, in particular, of changes in the climate and composition of the atmosphere. Additional studies will therefore be required.
7. **The impact of catch crops on the energy balance**. On the basis that a kilogram of nitrogen prevented from leaching corresponds to a reduction of the equivalent of around one kilogram of oil, reducing nitrate leaching into water via catch crops would lead to savings of several hundred thousand TOE (tonnes of oil equivalent) on a French scale. A precise evaluation of this issue would be useful in order to raise awareness among farmers and those who advise them of the energy-saving potential represented by this non-leached nitrate saving.
8. **Assessment of the impact of the widespread use of catch crops** in local situations in which groundwater recharge issues may be particularly important (case of groundwater resources replenished exclusively by drainage below crops). In this case, there may be a dilemma in terms of simultaneously addressing quantitative water management concerns (groundwater recharge) and qualitative objectives related to the protection of catchment water for human consumption (nitrate concentration of groundwater below the drinking water standard). This question will require specific studies, with simulations and combined agronomic and hydro-geological studies, along with explicit spatialization to take into account local characteristics related to the integrated operation of the

(continued)

agrosystem and the hydrosystem of groundwater. In addition, the long-term effects of routinely incorporating catch crops into French cropping systems is worthy of further research, especially with respect to carbon sequestration, the mineralization of nitrogen in the soil and the effects on pest and natural enemy populations.

9. **The design of decision-making tools to help select management techniques** for the implementation of catch crops that take into account relative local conditions in order to maximize the environmental benefits of catch crops and reduce their potential disadvantages. These local conditions are related to: climate and soil; cropping systems (particularly the previous and subsequent crops and soil tillage methods); labour organization on the farm. Decision-making tools of this type could help extend the use of catch crops in France.
10. **A cost-benefit analysis of cover crops**: a cost-benefit analysis of cover crops taking into account all of their functions would make it possible to better evaluate the services they provide, as well as to examine the most efficient government policies for their implementation.

All these questions form the framework for an ambitious research and R&D programme worthy of the millions of hectares of "Nitrate vulnerable zones" potentially concerned by the incorporation of catch crops into French cropping systems.

Erratum to: Cover Crops for Sustainable Farming

Eric Justes

Erratum to:
E. Justes (ed.), *Cover Crops for Sustainable Farming*, https://doi.org/10.1007/978-94-024-0986-4

The original version of this book was published with incorrect author names for chapters 1, 2 and 3. It is corrected and updated as:

Chapter 1. Study Context and Methodology
Authors: Eric Justes, Julie Constantin, Caroline Dürr, Carole Hermon, Christine Le Bas, Catherine Mignolet, Isabelle Savini, and Olivier Réchauchère

Chapter 2. Main Lessons Drawn from the Analysis of the Literature
Authors: Eric Justes, Nicolas Beaudoin, Patrick Bertuzzi, Raphaël Charles, Julie Constantin, Carolyne Dürr, Alexandre Joannon, Christine Le Bas, Bruno Mary, Françoise Montfort, Laurent Ruiz, Jean-Pierre Sarthou, Véronique Souchère, and Julien Tournebize

Chapter 3. Main Lessons Drawn from the Simulation Study
Authors: Eric Justes, Julie Constantin, Nicolas Beaudoin, Carolyne Dürr, Bruno Mary, and Laurent Ruiz

The original version of this book was published without a second sub-section in chapter 4. It is corrected and updated as:

4.2 Remaining Crucial Research Questions to be Explored

The updated online version of the original book can be found under
https://doi.org/10.1007/978-94-024-0986-4_1
https://doi.org/10.1007/978-94-024-0986-4_2
https://doi.org/10.1007/978-94-024-0986-4_3
https://doi.org/10.1007/978-94-024-0986-4_4
https://doi.org/10.1007/978-94-024-0986-4

E. Justes (ed.), *Cover Crops for Sustainable Farming*,
DOI 10.1007/978-94-024-0986-4_5

Bibliography

Cropping Practices and Regulatory Aspects

Arrêté du 6 mars (2001) relatif aux programmes d'action à mettre en oeuvre dans les zones vulnérables afin de réduire la pollution des eaux par les nitrates d'origine agricole
Arrêté du 7 mai (2012) relatif aux actions renforcées à mettre en oeuvre dans certaines zones ou parties de zones vulnérables en vue de la protection des eaux contre la pollution par les nitrates d'origine agricole
Chapelle-Barry C (2008) Enquête sur les pratiques culturales en 2006. Agreste chiffres et données, série agriculture 200
Circulaire du 26 mars (2008) relative aux modalités de mise en oeuvre des 4e programme d'action dans les zones vulnérables au titre de la directive n° 91/676 du 12 décembre 1991 concernant la protection des eaux contre la pollution par les nitrates à partir de sources agricoles, dite directive « nitrates »
Décret n° 2001-34 du 10 janvier 2001 relatif aux programmes d'action à mettre en oeuvre en vue de la protection des eaux contre la pollution par les nitrates d'origine agricole
Décret n° 2011-1257 du 10 octobre 2011 relatif aux programmes d'action à mettre en oeuvre en vue de la protection des eaux contre la pollution par les nitrates d'origine agricole
Directive 75/440 du 16 juin 1975 concernant la qualité des eaux superficielles destinées à la consommation humaine
Directive n° 2000/60/CE établissant un cadre pour une politique communautaire dans le domaine de l'eau
Directive n° 91/676 du 12 décembre 1991 relative à la lutte contre les nitrates d'origine agricole
Hermon C, Doussan I (2012) Production agricole et droit de l'environnement, Litec. LexisNexis, Paris, p 478
Loi n° 2006-1772 sur l'eau et les milieux aquatiques
Loi n° 2009-967 du 3 août 2009 de programmation relative à la mise en oeuvre du Grenelle de l'environnement
Rabaud V (2004) Peu de pratiques alternatives en grandes cultures. Agreste Primeur 149, p 4

E. Justes (ed.), *Cover Crops for Sustainable Farming*,
DOI 10.1007/978-94-024-0986-4

Cover Crop Management Techniques

Arvalis-Institut du végétal, Cetiom., ITB (2011) Cultures intermédiaires: impacts et conduite. Document de synthèse, p 236

Carrera LM, Abdul-Baki AA, Teasdale JR (2004) Cover crop management and weed suppression in no-tillage sweet corn production. Hortscience 39:1262–1266

Conseils pour le choix des couverts. La technique betteravière. Le betteravier français, n° 933, juillet 2010

Ekman S (2005) Cost-effective nitrogen leaching reduction as influenced by linkages between farmlevel decisions. Agric Econ 32:297–309

Hartwig NL, Ammon HU (2002) Cover crops and living mulches. Weed Sci 50:688–699

Hénin S, Gras R, Monnier G (1969) Le profil cultural. L'état physique du sol et ses conséquences agronomiques. 2e édition revue et corrigée, Masson et Cie. ed Masson, Paris

Hoffman ML, Regnier EE (2006) Contributions to weed suppression from cover crops. Food Products Press, Binghamton, pp 51–75

Joannon A, Papy F, Martin P, Souchère V (2005) Planning work constraints within farms to reduce runoff at catchment level. Agric Ecosyst Environ 111:13–20

Johnson WG, Davis VM, Kruger GR, Weller SC (2009) Influence of glyphosate-resistant cropping systems on weed species shifts and glyphosate-resistant weed populations. Eur J Agron 31:162–172

Martin P, Papy F, Souchère V, Capillon A (1998) Maîtrise du ruissellement et modélisation des pratiques de production. Cah Agric 7:111–119

Richard G, Boiffin J, Duval Y (1995) Direct drilling of sugar beet (Beta vulgaris L.) into a cover crop: effects on soil physical conditions and crop establishment. Soil Tillage Res 34:169–185

Smith AN, Reberg-Horton C, Place GT, Meijer AD, Arellano C, Mueller JP (2011) Rolled rye mulch for weed suppression in organic no-tillage soybeans. Weed Sci 59:224–231

Nitrate and Water During Fallow Periods

Allison MF, Armstrong MJ, Jaggard KW, Todd AD (1998) Integration of nitrate cover crops into sugarbeet (Beta vulgaris) rotations, I. Management and effectiveness of nitrate cover crops. J Agric Sci 130:53–60

Arlot MP (1999) Nitrates dans les eaux, drainage acteur, drainage témoin ? Thèse de l'université Paris VI, géologie appliquée, spécialité « hydrologie, hydrogéologie, géostatistiques et géochimie des eaux », p 374

Aronsson H, Stenberg M, Ulen B (2011) Leaching of N, P and glyphosate from two soils after herbicide treatment and incorporation of a ryegrass catch crop. Soil Use Manag 27(1):54–68

Askegaard M (2008) Residual effect and leaching of N and K in cropping systems with clover and ryegrass catch crops on a coarse sand. Agric Ecosyst Environ 123(1/3):99–108

Beaudoin N, Saad JK, Van Laethem C, Machet JM, Maucorps J, Mary B (2005) Nitrate leaching in intensive agriculture in Northern France: effect of farming practices, soils and crop rotations. Agric Ecosyst Environ 111:292–310

Berntsen J, Olesen JE, Petersen BM, Hansen EM (2006) Long-term fate of nitrogen uptake in catch crops. Eur J Agron 25(4):383–390

Cohan J.-P et al. (2011) Écophysiologie des couverts intermédiaires. Cultures intermédiaires : impacts et conduite. Arvalis-Institut du végétal, 236, 27–43

Constantin J, Mary B, Laurent F, Aubrion G, Fontaine A, Kerveillant P, Beaudoin N (2010a) Effects of catch crops, no till and reduced nitrogen fertilization on nitrogen leaching and balance in three long-term experiments. Agric Ecosyst Environ 135(4):268–278

Constantin J, Beaudoin N, Launay M, Duval J, Mary B (2011b) Long-term nitrogen dynamic in various catch crop scenarios: test and simulations with STICS model in temperate country. Agric Ecosyst Environ 147:36–46

Coppens P, Garnier P, De Gryze S, Merckx R, Recous S (2006) Soil moisture, carbon and nitrogen dynamics following incorporation and surface application of labelled crop residues in soil columns. Eur J Soil Sci 57:894–905

Dabney SM (1998) Cover crop impacts on watershed hydrology. J Soil Water Conserv 53:207–213

Dabney SM, Delgado JA, Reeves DW (2001) Using winter cover crops to improve soil and water quality. Commun Soil Sci Plant Anal 32(7–8):1221–1250

Davies DB, Garwood TWD, Rochford ADH (1996) Factors affecting nitrate leaching from a calcareous loam in East Anglia. J Agric Sci 126:75–86

Destain JP, Reuter V, Goffart JP (2010) Autumn cover crops and green manures: environment protection and agronomic interest. Biotechnol Agron Soc Environ 14:73–78

Di HJ, Cameron KC (2002) Nitrate leaching in temperate agrosystems: source, factors and mitigating strategies. Nutr Cycl Agroecosyst 46:237–256

Dinnes DL, Karlen DL, Jaynes DB, Kaspar TC, Hatfield JL, Colvin TS, Cambardella CA (2002) Nitrogen management strategies to reduce nitrate leaching in tile-drained midwestern soils. Agron J 94:153–171

Feaga JB, Selker JS, Dick RP, Hemphill DD (2010) Long-term nitrate leaching under vegetable production with cover crops in the Pacific Northwest. Soil Sci Soc Am J 74(1):186–195

Hansen EM, Djurhuus J (1996) Nitrate leaching as affected by long-term N fertilization on a coarse sand. Soil Use Manag 12(4):199–204

Justes E, Mary B, Nicolardot B (2009) Quantifying and modelling C and N mineralization kinetics of catch crop residues in soil: parameterization of the residue decomposition module of STICS model for mature and non mature residues. Plant Soil 325:171–185

Kaspar TC, Jaynes DB, Parkin TB, Moorman TB (2007) Rye cover crop and garnagrass strip effects on NO3 concentration and load in tile drainage. J Environ Qual 36:1503–1511

Machet JM, Laurent F, Chapot JY, Doré T, Dulout A (1997) Maîtrise de l'azote dans les intercultures et les jachères. In: Lemaire G, Nicolardot B (eds) Maîtrise de l'azote dans les agrosystèmes. Inra Éditions, Paris, p 271–288

Martinez J, Guiraud G (1990) A lysimeter study of the effects of a ryegrass catch crop, during a winter wheat/maize rotation, on nitrate leaching and on the following crop. J Soil Sci 41(1):5–16

Meisinger JJ, Hargrove WL, Mikkelsen RL, William JR, Benson VW (1991) Effects of cover crops on groundwater quality. In: Hargrove WL (eds) Cover crops for clean water, Jackson, Tennessee. Soil and Water Conservation Society, Ankeny, pp 57–68

Miguez FE, Bollero GA (2005) Review of corn yield response under winter cover cropping systems using meta-analytic methods. Crop Sci 45(6):2318–2329

Moller K, Stinner W, Leithold G (2008) Growth, composition, biological N2 fixation and nutrient uptake of a leguminous cover crop mixture and the effect of their removal on field nitrogen balances and nitrate leaching risk. Nutr Cycl Agroecosyst 82(3):233–249

Munoz-Carpena R, Ritter A, Bosch DD, Schaffer B, Potter TL (2008) Summer cover crop impacts on soil percolation and nitrogen leaching from a winter corn field. Agric Water Manag 95:633–644

Nicolardot B, Recous S, Mary B (2001) Simulation of C and N mineralisation during crop residue decomposition: a simple dynamic model based on the C:N ratio of the residues. Plant Soil 228:83–103

Shepherd MA, Webb J (1999) Effects of overwinter cover on nitrate loss and drainage from a sandy soil: consequences for water management? Soil Use Manag 15:109–116

Thomsen IK (2005) Nitrate leaching under spring barley is influenced by the presence of a ryegrass catch crop: results from a lysimeter experiment. Agric Ecosyst Environ 111:21–29

Thomsen IK, Christensen BT (2004) Yields of wheat and soil carbon and nitrogen contents following long-term incorporation of barley straw and ryegrass catch crops. Soil Use Manag 20:432–438

Thorup-Kristensen K, Dresboll DB (2010) Incorporation time of nitrogen catch crops influences the N effect for the succeeding crop. Soil Use Manag 26(1):27–35
Thorup-Kristensen K, Nielsen NE (1998) Modelling and measuring the effect of nitrogen catch crops on the nitrogen supply for succeeding crops. Plant Soil 203(1):79–89
Thorup-Kristensen K, Magid J, Jensen LS (2003) Catch crops and green manures as biological tools in nitrogen management in temperate zones. Adv Agron 79:227–302
Tonitto C, David MB, Drinkwater LE (2006) Replacing bare fallows with cover crops in fertilizerintensive cropping systems: a meta-analysis of crop yield and N dynamics Agriculture. Ecosyst Environ 112:58–72
Trinsoutrot I, Recous S, Bentz B, Linères M, Chèneby D, Nicolardot B (2000) Biochemical quality of crop residues and carbon and nitrogen mineralization kinetics under non limiting nitrogen conditions. Soil Sci Soc Am J 64:918–926
Unger PW, Vigil MF (1998) Cover crop effects on soil water relationships. J Soil Water Conserv 53:200–207
Zavattaro L, Monaco S, Saco D, Grignani C (2012) Options to reduce N loss maize in intensive cropping systems in Northern Italy. Agric Ecosyst Environ 147:24–35

Effects of Cover Crops on Erosion, Physical Soil Properties and Carbon Balance

Bodner G, Loiskandl W, Buchan G, Kaul HP (2008) Natural and management-induced dynamics of hydraulic conductivity along a cover-cropped field slope. Geoderma 146:317–325
Bodner G, Himmelbauer M, Loiskandl W, Kaul HP (2010) Improved evaluation of cover crop species by growth and root factors. Agron Sustain Dev 30:455–464
Campbell CA et al (2000) Organic C accumulation in soil over 30 years in semiarid southwestern Saskatchewan-Effect of crop rotations and fertilizers. Can J Soil Sci 80:179–192
Constantin J, Mary B et al (2010b) Effects of catch crops, no till and reduced nitrogen fertilization on nitrogen leaching and balance in three long-term experiments. Agric Ecosyst Environ 135:268–278
De Baets S, Poesen J, Meersmans J, Serlet L (2011) Cover crops and their erosion-reducing effects during concentrated flow erosion. Catena 85:237–244
Hermawan B, Bomke AA (1997) Effects of winter cover crops and successive spring tillage on soil aggregation. Soil Tillage Res 44:109–120
Joyce BA et al (2002) Infiltration and soil water storage under winter cover cropping in California's Sacramento Valley. Trans Asae 45:315–326
Kuo S, Jellum EJ (2000) Long-term winter cover cropping effects on corn (Zea mays L.) production and soil nitrogen availability. Biol Fertil Soil 31:470–477
Terzoudi CB, Gemtos TA, Danalatos NG, Argyrokastritis I (2007) Applicability of an empirical runoff estimation method in central Greece. Soil Tillage Res 92:198–212
Villamil MB, Bollero GA, Darmody RG, Simmons FW, Bullock DG (2006) No-till corn/soybean systems including winter cover crops: effects on soil properties. Soil Sci Soc Am J 70:1936–1944

Biotic Effects of Cover Crops on Weeds, Microflora and Fauna

Axelsen JA, Kristensen KT (2000) Collembola and mites in plots fertilised with different types of green manure. Pedobiologia 44:556–566

Besnard A, Duval R, Hopquin B, Lieven J, Morin P, Sträbler M (2011) Un choix d'espèces de plus en plus large. Cultures intermédiaires: impacts et conduite. Arvalis-Institut du végétal, Paris, 137–161

Blackshaw RE, Anderson RL, Lemerle D (2007) Cultural weed management. In: Upadhyaya MK, Blackshaw RE (eds), Non-chemical weed management: principles, concepts and technology. Cabi, Wallingford, pp 35–47. http://www.cabi.org/CABeBooks/default.aspx?site=107&page=45&LoadModule=PDFHier&BookID=378

Bonin L, Vacher C, Citron G, Pottier M, Lieven J (2011) Impact des couverts végétaux sur les adventices. Cultures intermédiaires: impacts et conduite. Arvalis-Institut du végétal, Paris, pp 78–84

Borges DC, Antedomenico SR, Santos VP, Inomoto MM (2009) Host suitability of Avena spp. genotypes to Meloidogyne incognita race 4. Trop Plant Pathol 34:24–28

Carreck NL, Williams IH (2002) Food for insect pollinators on farmland: insect visits to flowers of annual seed mixtures. J Insect Conserv 6:13–23

Chauhan BS, Gill GS, Preston C (2006) Tillage system effects on weed ecology, herbicide activity and persistence: a review. Aust J Exp Agric 46:1557–1570

Colbach N, Mézière D (2013) Using a sensitivity analysis of a weed dynamics model to develop sustainable cropping systems. I. Annual interactions between crop management techniques and biophysical field state variables. J Agric Sci 151:229–245

Colbach N, Granger S, Mézière D (2013) Using a sensitivity analysis of a weed dynamics model to develop sustainable cropping systems. II. Long-term effect of past crops and management techniques on weed infestation. J Agric Sci 151:247–267

Doré T, Sène M, Pellissier F, Gallet C (2004) Approche agronomique de l'allélopathie. Cah Agric 13:249–256

Ekeberg E, Riley HCF (1997) Tillage intensity effects on soil properties and crop yields in a long-term trial on morainic loam soil in southeast Norway. Soil Tillage Res 42:277–293

Fomsgaard IS (2006) Chemical ecology in wheat plant-pest interactions. How the use of modern techniques and a multidisciplinary approach can throw new light on a well-known phenomenon: allelopathy. J Agric Food Chem 54:987–990

Gallandt ER, Molloy T, Lynch RP, Drummond FA (2005) Effect of cover-cropping systems on invertebrate seed predation. Weed Sci 53:69–76

Heisler C, Brunotte J (1998) Assessment of tillage with plough and conservation tillage with regard to the biological activity by means of the bait-lamina-test after von Torne as well as the population density of springtails and predatory mites. Landbauforschung Volkenrode 48:78–85

Khanh TD, Chung MI, Xuan TD, Tawata S (2005) The exploitation of crop allelopathy in sustainable agricultural production. J Agron Crop Sci 191:172–184

Kirkegaard JA (2009) Biofumigation for plant disease control-from the fundamentals to the farming system. In: Walters D (ed) Disease control in crops: biological and environmentally friendly approaches. D. Wiley-Blackwell, Oxford, pp 172–195

Kirkegaard JA, Matthiessen J (2004) Developing and refining the biofumigation concept. Agroindustria 3:233–239

Kruidhof HM, Bastiaans L, Kropff MJ (2008) Ecological weed management by cover cropping: effects on weed growth in autumn and weed establishment in spring. Weed Res 48:492–502

Lankau RA, Wheeler E, Bennett AE, Strauss SY (2011) Plant-soil feedbacks contribute to an intransitive competitive network that promotes both genetic and species diversity. J Ecol 99:176–185

Liebman M, Davis AS (2000) Integration of soil, crop and weed management in low-external-input farming systems. Weed Res 40:27–47

Lu P, Gilardi G, Gullino ML, Garibaldi A (2010) Biofumigation with Brassica plants and its effect on the inoculum potential of Fusarium yellows of Brassica crops. Eur J Plant Pathol 126:387–402

Lundgren JG, Fergen JK (2010) The effects of a winter cover crop on Diabrotica virgifera (Coleoptera: Chrysomelidae) populations and beneficial arthropod communities in no-till maize. Environ Entomol 39:1816–1828

Mazzola M, Brown J, Izzo AD, Cohen MF (2007) Mechanism of action and efficacy of seed mealinduced pathogen suppression differ in a Brassicaceae species and time-dependent manner. Phytopathology 97:454–460

Motisi N, Doré T, Lucas P, Montfort F (2010) Dealing with the variability in biofumigation efficacy through an epidemiological framework. Soil Biol Biochem 42:2044–2057

Ortiz-Ceballos AI, Fragoso C, Brown GG (2007) Synergistic effect of a tropical earthworm Balanteodrilus pearsei and velvetbean Mucuna pruriens var. utilis on maize growth and crop production. Appl Soil Ecol 35:356–362

Prasifka JR, Schmidt NP, Kohler KA, O'Neal ME, Hellmich RL, Singer JW (2006) Effects of living mulches on predator abundance and sentinel prey in a corn-soybean-forage rotation. Environ Entomol 35:1423–1431

Pyrowolakis A, Schuster RP, Sikora RA (1999) Effect of cropping pattern and green manure on the antagonistic potential and the diversity of egg pathogenic fungi in fields with Heterodera schachtii infection. Nematology 1:165–171

Reau R, Bodet J-M, Bordes J-P, Doré T, Ennaifar S, Moussart A, Nicolardot B, Pellerin S, Plenchette C, Quinsac A, Sausse C, Seguin B, Tivoli B (2005) A review on brassicas allelopathic effects through their interaction with soil borne pathogens and mycorrhizas. Part 1. OCL-oleagineux Corps Gras Lipides 12:261–271

Robin N et al. (2011) Impact des couverts végétaux sur les ravageurs. Cultures intermédiaires: impacts et conduite, Arvalis-Institut du végétal, Paris, pp 91–98

Teasdale JR (1996) Contribution of cover crops to weed management in sustainable agricultural systems. J Prod Agric 9:475–479

Teasdale J, Daughtry CST (1993) Weed suppression by live and dessicated hairy vetch (Vicia villosa). Weed Sci 41:207–212

Teasdale J, Mohler CL (2000) The quantitative relationship between weed emergence and the physical properties of mulches. Weed Sci 48:385–392

Teasdale J, Brandsaeter LO, Calegari A, Skora-Neto F (2007) Cover crops and weed management. In: Upadhyaya MK, Blackshaw RE (eds) Non-chemical weed management: principles, concepts and technology. Cabi, Wallingford, pp 9–64

Thoden TC, Boppre M (2010) Plants producing pyrrolizidine alkaloids: sustainable tools for nematode management? Nematology 12:1–24

Wang KH, McSorley R, Marshall A, Gallaher RN (2006) Influence of organic Crotalaria juncea hay and ammonium nitrate fertilizers on soil nematode communities. Appl Soil Ecol 31:186–198

Wiggins E, Kinkel LL (2005) Green manures and crop sequences influence alfalfa root rot and pathogen inhibitory activity among soil-borne streptomycetes. Plant Soil 268:271–283

Methodology of the Simulation Study

Brisson N, Ruget F, Gate P, Lorgeau J, Nicoullaud B, Tayot X, Plenet D, Jeuffroy M-H, Bouthier A, Ripoche D, Mary B, Justes E (2002) STICS: a generic model for simulating crops and their water and nitrogen balances II. Model validation for wheat and maize. Agronomie 22:69–92

Brisson N, Gary C, Justes E, Roche R, Mary B, Ripoche D, Zimmer D, Sierra J, Bertuzzi P, Burger P, Bussiere F, Cabidoche YM, Cellier P, Debaeke P, Gaudillere J-P, Henault C, Maraux F, Seguin B, Sinoquet H (2003) An overview of the crop model STICS. Eur J Agron 18:309–332

Dorsainvil F, Dürr C, Justes E, Carrera A (2005) Characterisation and modelling of white mustard (Sinapisalba L.) emergence under several sowing conditions. Eur J Agron 23:146–158

Inra (1998) Base de données géographiques des sols de France à l'échelle du 1/1 000 000, version1 du 21/12/1998. Inra, US 1106 Infosol, Orléans, France

Lefebvre MP (2010) Spatialisation de modèles de fonctionnement hydromécanique des sols appliquée à la prévision des risques de tassement à l'échelle de la France. Thèse université Orléans, p 295

Marcuola F, Mignolet C, Schott C, Mari JF (2010) Organisation spatiale des successions culturales en France entre 1992 et 2009. Projet ANR-POPSY p 39

Richard G (coord.) (2008) Projet DST. Dégradation physique des sols agricoles et forestiers liée au tassement.: impact, prévision, prévention, suivi, cartogtaphie, Inra Science du sol, Orléans, p 30

Authors

Experts authors

Alexandre JOANNON, Inra, Sciences pour l'action et le développement, Rennes, France, unité Paysages

Bruno MARY, Inra, Environnement et agriculture, Laon, France, unité Agrosystèmes et impacts environnementaux carbone-azote

Carole HERMON, université Toulouse 1 Capitole, France

Carolyne DÜRR, Inra, Environnement et agronomie, Angers, France, unité Physiologie moléculaire des semences

Catherine MIGNOLET, Inra, Sciences pour l'action et le développement, Mirecourt, France, unité Agrosytèmes territoires ressources, Mirecourt

Christine LE BAS, Inra, Environnement et agriculture, Orléans, France, unité Infosol

Éric JUSTES (editor), Inra, Environnement et agronomie, Toulouse, France, unité Agrosystèmes et agricultures, gestion de ressources, innovation et ruralité, responsable scientifique de l'étude

Françoise MONTFORT, Inra, Santé des plantes et environnement, Rennes, France, Institut de génétique environnement et protection des plantes

Jean-Pierre SARTHOU, Ensat Toulouse, France, unité Agrosystèmes et agricultures, gestion de ressources, innovation et ruralité

Julie CONSTANTIN, Inra, Environnement et agronomie, Toulouse, France, unité Agrosystèmes et agricultures, gestion de ressources, –innovation et ruralité

Julien TOURNEBIZE, Irstea, Antony, France, unité Hydrosystèmes et bioprocédés

Laurent RUIZ, Inra, Environnement et agriculture, Rennes, France, unité Sol agro et hydrosystème spatialisation

Nicolas BEAUDOIN, Inra, Environnement et agronomie, Laon, France, unité Agrosystèmes et impacts environnementaux carbone-azote

Patrick BERTUZZI, Inra, Environnement et agronomie, Avignon, France, unité Agroclim

Punctual contribution of Thomas HENRY, Trainee engineering school ofPurpan, France

Raphaël CHARLES, station de recherche Agroscope Changins-Wädenswil ACW, Suisse

Véronique SOUCHÈRE, Inra, Sciences pour l'action et le développement, Grignon, France, unité Activités, produits, territoires

Driving of the Project

Hugues LEISER, Inra, Avignon, France

Isabelle SAVINI, Inra Depe, Paris, France

Marion BARBIER, Inra Depe, Paris, France

Olivier RÉCHAUCHÈRE (coordinator), Inra Depe, Paris, France

MIX
Papier | Fördert gute Waldnutzung
FSC® C083411

Kolibri 360 GmbH
Ferdinand-Jühlke-Straße 7
99095 Erfurt, Deutschland
produktsicherheit@kolibri360.de